THAT'S THE WAY THE BALL BOUNCES

THAT'S THE WAY
THE BALL BOUNCES

❖

A Nonplused Soldier's Mundane Exploits
During WW II

MILTON COOK

Copyright © 2008 by Milton Cook.

Library of Congress Control Number: 2007907571
ISBN: Hardcover   978-1-4257-9867-3
               Softcover   978-1-4257-9853-6

All rights reserved. No part of this book may be reproduced or transmitted in any form or by any means, electronic or mechanical, including photocopying, recording, or by any information storage and retrieval system, without permission in writing from the copyright owner.

This book was printed in the United States of America.

To order additional copies of this book, contact:
Xlibris Corporation
1-888-795-4274
www.Xlibris.com
Orders@Xlibris.com

41791

Contents

Acknowledgments 15
Forward 17
List of Photographs

1257 Military Police Company (AVN) 56
Labor In Vain Inn 66
Station Periphery Fence 68
MP ME Cook 78
1. Front of MP Barrack 79
2. Co-inductees 79
3. Rows of Barracks 79
4. View from Right Wing MP Barrack 79
5. Permanent Party Officers Mess & MH 80
6. Permanent Party Officers Mess & Other 80
A Post Exchange 81
American Red Cross Canteen 91
High Street 91
Eighth Air Force Shoulder Patches 101
A Chow Line 105
The Potteris 115
Duncan Hall Entrance (No 2) Gate House 137
Hanley Red Cross Club 143
View of Street From Red Cross Window 146
Changing of the Guard—Oct 1945 146
Battle of Britain Parade, Eccleshall 157
Battle of Britain Parade Stafford 160
Bill Amrhein 164
Howard Hall Theater 168
An "Enlisted Men's" Snack Bar 170

Duncan Hall Entrance (No 2) Gate	190
Colonel Rader, General Spaatz, General Doolittle	194
Soldiers Three—Cook, Foiles, Streialaw	196
Off Season Holiday in Blackpool	198
Nelson Hall	202
Nelson Hall From the Air	202
Duncan Hall (No. 1) Exit Gate	205
The Motor Pool	210
The Main Headquarters	217
Colonel Rader Reviews the 695th AAF Band and The MP Co	229
MP Company on Parade	229
Duncan Hall Dispensary	231
Ready for Inspection	231
1257 Military Police Company (AVN)	242
Chateau Rothschild	256
Paris 1945	260
Military Police Off Duty	273
Military Police Ready For And On Duty	274
First Billet in Large Room	297
The Stair Case	298
The Hallway	298
Headquarters Building and Relief	303
At Leisure On The Grounds	303
Cook and Szucs in Action	304
At Ease	314
Camp Lucky Strike	317
A Nice Day on The North Atlantic	320
New York Harbor	322
Camp Kilmer	324
Back in California	325
Johnson	344
Reardon	344
Vernon	344
Postcard Street Staffs	357
Postcard Square Staffs	358
Armed Forces Book Notice	366

List of Documents

Graduation Program	23
Draft Notice	28
Life Insurance Policy	33
Pay Record Booklet	47
PX Ration Card	82
US Army ID Card	83
Class "B" Pass	84
V-Mail	139
Two-Day Pass to London	145
Battle of Britain Day Service	158
Three-Day Pass	327
Hurlstone Letter	359
Special Orders No. 70	378
Train Schedule for Special Orders No. 70	379
Special Orders No. 6	380
Train Schedule for Special Orders No. 6	381
Special Orders No. 66	382
Special Orders No. 78	383
Train Schedule for Special Orders No. 78	384
Special Orders No. 82	385
Security Clearance for Special Orders No. 82	386
Special Orders No. 38	387
Special Orders No. 68	388
Class "B" Pass	390
Mess Pass	391
A Two-Day Pass	392
A One-Day Pass	393
Last PX Ration Card	394
Class "B" Pass	395

Part I—The Fringe of History

1. Ambiguous Interlude 23
2. Caught in the Draft 27

Part II—Teenagers With Guns

3. Introduction to the Army 31
4. A Trip to Florida 35
5. Basic Training 37
6. Formation of a Company and Additional Training 44
7. Trip to Minnesota 48

Part III—An Atypical Assignment

8. Camp Miles Standish, Massachusetts 59
9. Across the Atlantic 62
10. The Trip to Stone 65
11. A Trip North to Chorley 71
12. Back South to Yarnfield 77
13. Required Credentials 81
14. Settling In For Multiplex Duty 85
15. Rambling on Foot Patrol 87
16. Rambling on Patrol-Stafford 89
17. Rambling on Patrol-Stone 90
18. Rambling on Patrol-Eccleshall 93
19. Rambling on Patrol-Millmeece 97
20. Rambling on Patrol-Back At the Guardhouse 98
21. The Way of The WACs 99
22. Occurrences In the Barrack-Best Foot Forward 100
23. Dining in the Mess Hall-Holiday Feasts 105
24. Dining in the Mess Hall-Some Good Food
    and Some Bad 106
25. Singular Service-Left Out in the Cold 108
26. Tales From the Guardhouse-Love and War 111
27. Singular Service-Orders to East Anglia 116
28. Tales from the Barrack-Cleaning Detail 128
29. Guarding the Gate-Always on the Watch 130
30. Singular Service-Point Control 133
31. Guarding the Gate-A Woman of Mystery 135
32. Tales from the Barrack-Packages from Home 138

33. Off Duty-Passing It Around 142
34. Guarding the Gate-A Rueful Account............................... 149
35. Guarding the Gate-Reunions... 151
36. Singular Service-Meeting With a P-38............................... 153
37. Battle of Britain Parades 156
38. Tales from the Barrack-Crossing Paths 161
39. A Singular Comrade... 163
40. Singular Service-Tour of the Town.................................... 165
41. Entertainment-The Theater 167
42. Singular Service-Orders to London Town 171
43. Tales from the Guardhouse-Suicide Watch 182
44. Singular Service-Time Out With a P-47 185
45. Entertainment-The P.A. System... 188
46. Guarding the Gate-A Generally Fine Day 191
47. Entertainment-A Furlough At Last 195
48. Singular Service-Guarding the Gate at Nelson Hall 200
49. Occurrence in the Barrack-A GI Party.............................. 203
50. Guarding the Gate-Cold Feet Again 205
51. Tales From the Barrack-Troop Movements 209
52. Tales From the Barrack-Buy U.S. War Bonds 213
53. Singular Service-Orders to Chorley 215
54. Guarding the Gate-Action in the STO
    (Social Theater of Operation) 225
55. Singular Service-Orders to East Anglia Again 232
56. Tales from the Barrack-Company Picture........................ 241
57. Venerable Edifices Then
    and Again-Westminster Abbey.. 243
58. Venerable Edifices Then
    and Again-The Tower of London....................................... 246
59. Singular Service-Orders to Paris.. 249
60. A View From the Barrack-VE Day..................................... 266
61. Occurrences in the Barrack-A Party of GIs..................... 268
62. Tales From the Barrack-The Good Old Army Life............... 272
63. Tales From the Guardhouse-Among the Dishonored Dead..... 275
64. Singular Service-Orders to London Town Twice Again....... 280
65. Departing From the Barrack-Good Bye Howard Hall 284

Part IV—Army of Occupation

66. Off to Germany 289
67. A Jaunt Through the Town.. 294
68. Life in the Barracks 296
69. GI Cough Drops 300
70. Boys Will Be Boys 302
71. Guarding the Gate 'Till the Very End................................ 307

Part V—Heading Home

72. The Last Time I Saw Paris.. 313
73. I Won't Be Home For Christmas 316
74. Going Home-Tempest Tossed.. 319
75. Going Home-Turbulence Tossed 323

Part VI—Back Home Again

76. Lost in the Desert... 331

Part VII—The Denouement

77. The American Soldier in World War II.............................. 341

End Notes 351
Appendix I—Special Orders and Travel Documents 377
Appendix II—Other Documents 389
Bibliography 397

For Roberta

Who would be both pleased and proud,
but who I think, might well have been taken aback
by some of the things of which I tell.

< NOTICE >

WELCOME TO WORLD WAR II
THREE (3) THEATERS TO CHOOSE FROM
ENJOY THE SHOW !
HAVE FUN !
YOU SHOULD BE AT LEAST
EIGHTEEN (18) YEARS
OLD TO ATTEND.

> WARNING <

SOME ACTIVITIES MAY BE
DETRIMENTAL TO YOUR HEALTH,
MAY RESULT IN PHYSICAL AND IN
EMOTIONAL DISTRESS AND IN SOME
CASES EVEN YOUR DEATH.

| IN ANY CASE IT WILL BE MOSTLY BORING. |

NOT YOU NOR ANY ONE ELSE WILL
HAVE ANY CONTROL OVER EVENTS
AS THEY OCCURE.

| BUT THAT'S THE WAY THE BALL BOUNCES! |

ACKNOWLEDGMENTS

There was a time when I could type not to fast but reasonably well. However, working for many years as a mechanical engineer, even up to being a chief engineer, most of what I wrote by hand was transcribed to type by a "secretary." This was the way much text was prepared before the advent of the personal computer. After so many years of writing reports, plans, procedures, technical papers and uncountable letters in long hand—actually I printed—my brain is programmed to put it down on paper by hand. And so it was with this memoir; after several hand-printed drafts my daughter Linda Alvarez would transcribe my more or less legible hand written text in to print using Microsoft Word. This procedure continued until about two-thirds of the memoir had been written. And then for the rest of this chronicle I did my own transcribing using Dragon Naturally Speaking.

I have a considerable collection of photographs of fellow soldiers and of some of the facilities where we were for a short or for a long time. But then to supplement those pictures from my own collection I shamelessly excerpted photographs from *The Yarnfield Yank*. This post war publication was a collection of pictures assembled from the Replacement Depot's photo archives. So, to save some time I used some of the photos from this collection instead of locating and reviewing what must be a very large file in the National Archives.

In the endnotes I have included the complete article, from what was probably the last issue of the stations post war newspaper, the *Ardee News*. This article, *Impressive Figures in Command Record*, by Dave Clark, provides a concise record of the deployment and repatriation of Army Air Force soldiers in Briton and in Europe by the replacement depots, during and after the war.

Otherwise what I have written in this account is from my memory supplemented by the documents that I had saved and by an occasional reminder of some things or events that are described in one of the current histories of the American soldier in England during World War II.

FORWARD

Histories are replete with errors both important and trifling that stem from flawed data or imperfect recollections and this memoir is no exception. It is a mostly true chronicle of a different time and a different style of life, as it was so very many—many worlds ago. For a long time I had mulled over the idea of recounting some of the unusual experiences that I had while serving in the United States Army Air Force during Second World War. With my wife's passing I concluded that it was now time to get on with the show and relate some of the particular events and goings-on with which I had been involved. I thought it was important that some of the not commonly known aspects of the U.S. Army's operations in England, which I remembered and certainly my own some times uncommon experiences as a military policeman should be recounted. The events in which I was entailed while sometimes interesting, where for the most part not very exciting, nor were they I think, very significant in the grand endeavor of the U.S. Army's mission in the ETO. And never during the whole time spent in the war zone was I ever in any immediately apparent harms way. With the passage of time the accumulated knowledge, gained from the "war stories" of other former soldiers experiences of "normal" Army life, I became even more persuaded that many of my personal experiences were very much out of the ordinary. Its possible that this account of my service in the U.S. Army Air Force during the Second World War will provide at least a footnote to that history of the war. This history does treat some aspects of the Army's activities during and after the war was over, which to my knowledge no one else is probably aware of and in fact may be not even have considered.

Initially my knowledge of the recorded historical minutia of the U.S. Army Air Force in England during WW II was very limited. But as I proceeded with this particular tale I discovered that there were a number of available books devoted to the U.S. Army's presence and activities in the British Isles. I found Reynolds' significant history *Rich Relations* very helpful as it clarified some circumstances and refreshed my memory of items that I had not initially thought to write about. Freeman's books on the Mighty Eighth Air Force and Bowman's *Wild Blue Yonder* were also sources of pertinent information and memory refreshment. A result of reviewing these books and other "back ground" material my personal story got out of hand and developed in to a semi-scholarly history.

Books are still being written about the American Civil War so it is likely that the occupation in Britain during World War II is a lode that will be mined in detail for many years to come. And in that view some of the subjects that I have briefly touched upon could be examined more completely than was necessary in this account of my stay in England. Many nit-picking details have sometimes been included in this narrative as it is the common things, of which everyone knows, that are very seldom included in historical accounts. It may be that some of this detailed information will keep some future historian from speculating: just what was the function of this particular artifact or just what was the meaning of a strange expression.

In any case my memory of events has been somewhat degraded by the passage of time. Some of the events I am certain did not occur as they have been recounted, but I think that at least the substance has been related. Many accounts have been aided or based on the papers and the photographs that I have saved; but which would be of little significance to anyone other than myself. As much as I could I have tried to provide some historical context for the various documents and to elucidate my individual view of the war. In too many cases I can no longer match up the names with the faces in the photographs. Some of my observations are based on the review of incidents, considering later experience and historical records, to provide some insight into these events and these times. This memoir is my particular and peculiar history and my take on events including the Army culture of that time. Included in this accounting are some details of the stations, dispersed about England and later in France, which were responsible for the assignment, the combat rest and the repatriation of the U.S. Army Air Force personnel in England and later in northern Europe. My service in England was as a member of the 93rd Station Complement, the headquarters unit of the distributed personnel system. And even more so this is some of the history of one unit of the 93rd Station Complement, that unique coterie the 1257th Military Police Company (AVN). Still having the names of some sixty who were members of the 1257th I have used them somewhat indiscriminately except for those instances where I do remember who it was and when it was. Overall, I have aimed to reflect the attitude and the mores of that different time and that different culture.

This is a story that emerges through the accrual of the details of what were for the most part trivial events. So now come along with me as I takeoff after my own distant drummer beating away on the crest of the hill, and what a strange cadence she keeps.

PART I
THE FRINGE OF HISTORY

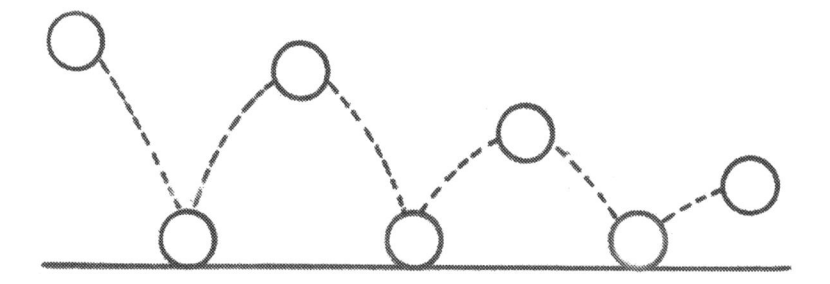

CHAPTER 1

Ambiguous Interlude

> Twenty-Sixth Annual
> Commencement
>
> VICTOR VALLEY
> HIGH SCHOOL
>
> THURSDAY EVENING, JUNE FOURTH
> NINETEEN HUNDRED FORTY-TWO

    I was still seventeen and this commencement was the very last act of my high school student days. There were forty of us exactly twenty boys and twenty girls, in the class of '42 who were graduating from the Victor Valley Union high school on that capping night in June. We had no inkling then or any knowledge for a long time to come that the tide of the war with Japan was turning at that very point in time. For, the Battle of Midway had commenced and Japan was to suffer the first of a series of defeats. Ironically the Battle of Midway, which resulted in the unrecoverable loss of aircraft carriers for the Japanese, was the consequence of the resolve by Admiral Yamamoto to destroy America's aircraft carriers. He had ordered a Japanese

23

naval task force to locate, engage and extirpate the U.S. Navy Pacific Fleet aircraft carriers, and thus eliminate the risk of any more assaults on the islands of Japan. Attacks such as the Tokyo air raid which had come about as a result of a chance remark that B-25 Bombers were capable of being launched from the flight deck of an aircraft carrier. Yamamoto desired to avoid any further embarrassment such as the bombing raid on Tokyo led by the then the Lieutenant Colonel Doolittle. Who was a soon to be general with who, in the future, I would cross paths very quickly.

There are quite a few of my classmates whom I never saw again after that night. Some of my ex-classmates I have occasionally seen, and some I have stayed in touch with over the years, as fate has allowed. There are a few of my former classmates with whom I have spent varying amounts of time over the years.

Pass on Class of '42!

00000

While attending high school, I had worked a couple of hours after school and on Saturdays for an egg farmer. With the limited income from this job, I was able to cover all school expenses such as supplies, lunches and clothing. At graduation, I had saved money, which would take care of my petty need for a while. However, I did not have enough cash to last me until I was eighteen and could obtain employment in a defense factory. I knew full well that any job that I would find at this time, because of my age and the war, would only be temporary.

My mother and my father had been legally separated for some time. My father was wintering in Bodfish, California. He was about to relocate back up on Mount Piute in the Tehachapi Mountain Range. My father was operating an old gold mine, which was owned by the Morelands, who operated the adjoining gold mine and who lived at the mine full time. I decided that I would go with my father, my two younger brothers and to the gold mine for the summer. I intended to return to Adelanto in September for my birthday, after which I would be qualified for regular employment.

I helped my father with his operation of the gold mine and spent some time exploring the surrounding forest. There was evidence of large-scale gold mining and prospecting that had occurred in the past. There were a lot of interesting artifacts to see and features to discover. One day as I was walking along a trail, I came face to face with a large cinnamon bear coming the other way. As soon as we spotted each other, we each turned tail and returned back down our respective portion of the trail.

In the first part of September, my father drove me back to Adelanto. After my eighteenth birthday, I found a job washing pots and pans in the kitchen of a mess hall on George Air Force Base. The George Air Force Base was located between Adelanto and the Mojave River. I had hiked to the Mojave River many times where the Air Force Base was now situated. In fact construction had been started even before the U.S. had "officially" entered the war. Several friends and I hiked out to the site on December 7, 1941 to see what progress had been made. And there in the sandy desert, the brush still not entirely cleared away, we saw the freshly poured concrete of the main runway. The end of the brand new runway, which we looked at on that Sunday morning, would later have to be replaced as it could not sustain the repeated pounding of the landing of large airplanes. I spent time with the pots and pans for just one month, and then I collected my pay and headed for the big city, Los Angeles.

Specifically, I was headed for a rooming house where my high school buddy, Bob Vernon, was living. A rooming house operated by the mother-in-law of Bob's brother Bill (who was then in the U.S. Navy). This rooming house was situated a few blocks from Glendale Boulevard, where Los Angeles protrudes up between Griffith Park and Glendale. The right-of-way for the Red Car tracks ran down an esplanade in the middle of Glendale Boulevard. The Red Car stop was a short walk from the rooming house. The Red Car was the first leg of my friend, Bob's commute to Douglas Aircraft at Clover Field in Santa Monica where he was working on the swing shift

I had scheduled an employment interview at Lockheed Aircraft in Burbank. It would only be a short commute from the rooming house to Burbank. The night before my interview was to take place President Roosevelt recommended on October 12 in a radio address "fireside chat" that the Selective Service should start drafting eighteen and nineteen-year-olds. The next morning when I arrived for my interview at Lockheed Aircraft, I was told, "You are going to be drafted soon, and we are not going to talk to you or hire you. Good bye." On November 18, Congress amended the Selective Service Act to lower the draft age to eighteen years.

One day while riding to Douglas Aircraft on the streetcar, I had a chance encounter with one of my high school classmates, Dorothy Harr. I had sat next to her in one class and gave her a hard time about the huaraches, which she was always wearing. I told her that they stank like a wet corral, and they did. This chance encounter was the last time that I ever saw her.

The first job assigned to me at Douglas Aircraft was as a Parts Treater. In this function, my main duty was to dip large wire baskets of aluminum casting in a very hot salt bath. After a while on this job, I was reassigned to an outside area where we leak tested large steel tanks. These tanks were to be used to provide extra fuel for aircraft that were to be flown to Russia via Alaska and the Bearing Straits. The leak testing of the fuel tanks was completed, and I was transferred to a job on the swing shift. This new job title was Hand Forming, and it consisted of using an arbor press to make slight corrections to the shape of formed aluminum parts. The aluminum parts had been heat-treated and were kept very cold in a walk-in refrigerator. I removed one part at a time from the refrigerator, because they had to be formed while they were cold as they became hard at room temperature and could no longer be bent.

On the swing shift, the order of the commute from the rooming house to Santa Monica did not change. However, late at night when the swing shift ended, the first inbound yellow car was not operating. Bob and I would ride what was the second yellow car all of the way to downtown L.A. We then walked a few blocks to the subway terminal, which I think was located at Hill and 5th Street. The subway terminal was the end of the line for the Red Car, which we had to take to get home. There was a long ramp down to the Red Car boarding platform for the cars. The Red Cars ran through a tunnel under Bunker Hill. To this day, you can still see the remains of the roadbed for the Red Car tracks on the hillside of the Golden State (5) Freeway.

My friend, Bob and I had a carefree time on weekends hitting the bookstores and motion picture theaters in downtown LA and in Hollywood. Smoking was allowed in the loge section of the movie theaters in those days. We were real sophisticates and would smoke our "Anthony & Cleopatra" cigars as we watched the movie. It was Christmas Day 1942, I was smoking away and enjoying the movie, and it hit me that this was really stupid. And so I quit smoking forever.

CHAPTER 2

Caught in the Draft

Sometime near the end of the year, Bob and I registered for the Draft. On a Saturday, we walked to the local office of the Draft Board. The office was located just over the bridge, as it exists today, on which Glendale Boulevard spans the Golden State Freeway. Then in February, there came our "greetings." That's what the Draft Notice said, "Greetings, you have been selected by your friends and neighbors . . ." but we had been selected to fill a quota. Our neighbors had no idea just who we were, and could not possibly be our friends. The Draft Notices instructed us to report to the Los Angeles Induction Center on March 8, 1943.

A peculiar practice, the draft: a requisition for an individual to spend some portion of their life or to maybe even expend it in the military service of their country. Enlistments in to the military service tend to waver and decline after the first flush of euphoria that comes with a brand new war, as many believe "maybe just not yet". To make up for the shortfall of enlistments it has been the social custom, probably since the inception of "civilization", to press young men in to the military to satisfy the exigency of war. So, it always has been an unambiguous duty to serve one's country when called upon, not withstanding that to not do so would result in one's imprisonment at the very least.

DRAFT NOTICE

An order from the President of the United States, but which was however issued by a local board of the Selective Service System to report for induction in to the Armed Forces. What a country this is where the leader himself takes an interest in the conscription of an ordinary citizen in to the military organization of the nation.

PART
II
TEENAGERS WITH GUNS

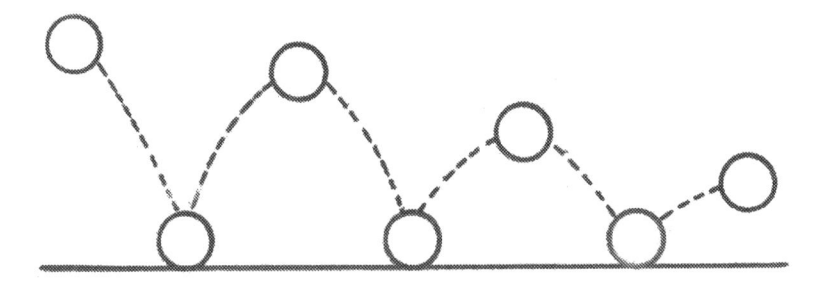

CHAPTER 3

Introduction to the Army

Early in the morning on the eighth of March in 1943, Vernon and I arrived at the Military Induction Center in downtown Los Angeles. After check-in, along with the other draftees were required to complete some forms. After finishing his paper work, each draftee was subjected to a rudimentary medical examination. Those of the draftees who had passed the medical examination were required to take an oath to uphold and defend the constitution of the United States of America, and were thus sworn into military service. We inductees were then loaded on to a bus to be transported to an army base located in Arlington, California.

In addition to serving as an induction center, the Arlington Army Base involved other Army activities, as it apparently was also a supply depot. We inductees were members of the Army of the United States, while those who had enlisted were members of the United States Army. Except for this slight technicality, we were, for practical purposes, all in the U.S. Army. All of the members of the Army who did not have a commission or a warrant were referred generally as enlisted men—more commonly, the designation was simply "EM." I wonder, were the enlisted women referred to as "EW?" At the time, the designation "EM" did not bother me, but subsequently I changed my mind. So, I strive through out this narrative to avoid denoting myself as "enlisted."

The first order of business was to issue new uniforms and other necessary paraphernalia to all of us inductees. We were assigned quarters and instructed to change into our new uniforms and at least look somewhat like soldiers. We then packaged up all of our civilian clothing for shipment to our former homes. This would be the very last that we would ever see of our civvies, because for the most part our younger siblings would soon be wearing them. And what the heck they would most likely be out of style by the time that we returned home.

Part of the accouterments was our bedding, which consisted of two olive drab (OD) blankets and mattress covers. The mattress cover was always referred to as a "fart sack." The mattress cover was also used as a burial shroud. Using this bedding, we then made up our bunks. All of the rest of our equipage including spare uniforms was packed into our A and B barracks bags. Remarkably, the short time at the Arlington Army Base would be the last time I would be quartered in a standard Army barrack until almost three years later when I returned to the United States to be discharge from the Army.

The government issued (GI) each of us three types of uniforms. The fatigues were light green cotton denim for work and casual wear. For the summer dress uniform, we were provided with cotton khakis, which were referred to as "suntans." The summer dress uniform included khaki tie, garrison cap and web belt. The OD winter uniform, in addition to wool shirt and pants, included a wool service jacket with lacquer coated dull brass buttons and a very heavy overcoat with olive green plastic buttons. A wool OD garrison cap was part of the winter uniform along with the khaki tie and belt. Other parts of the uniforms included a cotton poplin field jacket, brown high top shoes and dismounted leggings. The service jacket, which I was issued, was way too large. I said to the supply Sergeant, "Hey, this jacket is too big." To which he replied, "You'll grow into it." I never did grow into the service jacket, and after I arrived in England, I had it tailored to fit me.

The next morning, all of the inductees were subjected to intelligence and aptitude testing. Then, each of us was interviewed to determine where we would most "benefit" the Army. My buddy, Bob Vernon was determined to have an aptitude for Morris Code and was sent to the Army Air Corps Radioman School. Bob ended up flying thirty missions over Germany as a waist gunner on a B-24. As for myself, I requested that I be assigned to the Army Corps of Engineers, as I wanted to blow things up. However, it is just as well that I did not end up in the Army Engineers, as I might have blown myself up as well. Anyway, I was told, "No, you are assigned to the Corps of Military Police. You are big, and you are intelligent, and that is what is needed in the Military Police. The U.S. Army had lots of trouble in World War I because the 'big, dumb guys' were put into the Military Police. The Army was going to do it right this time and put people who could think in the MPs." However, it's my theory that there was a quota for the MPs that I helped to fill. Evidence for the Quota Theory is found in the fact that two of my fellow inductees, James Collins and Donald Foiles, were also placed in the MPs. In fact, the three of us were in the same outfit until after the end of the war. We were advised to sign up for the National Service Life Insurance. This was a term insurance policy underwritten by the Veteran's Administration, as the private sector insurance companies were reluctant to cover Armed Forces personnel who would more than likely soon enough be in harms way. But, the risk was not very significant, because it would turn out that over ninety seven percent of those who served in the military forces would survive World War II. I selected the $10,000 policy, which at the time was a substantial amount. The payment for this life insurance policy was $6.40 per month that would be deducted from my pay, as an Army Air Force Buck Private, of $50 per month.

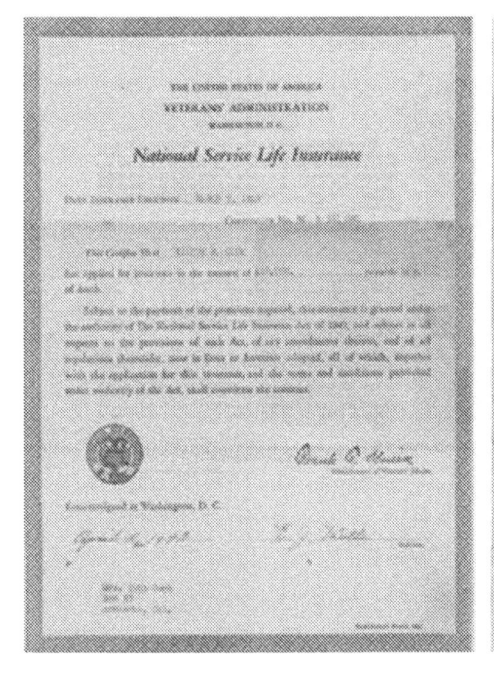

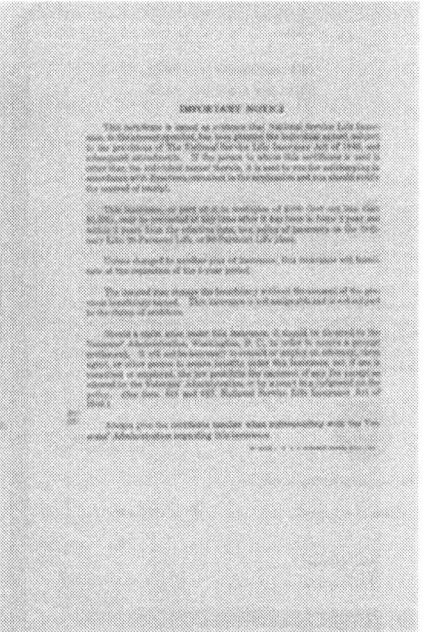

LIFE INSURANCE POLICY

All of us inductees were given shots for tetanus, typhoid and, for those who had never had one, a small pox vaccination. Next, it was new haircuts for all. The Army also presented a gift of two brass "dog tags" on a bead chain to wear around our necks at all times. We were encouraged to learn, and recite when called upon, our serial number that was impressed into our dog tags.

I had watched as a clerk incised the dog tags. Using a typewriter like keyboard he imprinted the dog tags with the soldiers name, serial number, T-43 (year of first tetanus shot), blood type, next of kin name and address, and a single letter religion identifier. The thin brass blanks, with folded over double thick edges, were one-inch wide by two inches long with rounded ends. A one-eighth inch hole near the edge at the center of the right end was somewhat complimented by a same size rounded notch in the upper arc of the other end. The blanks were held in place on the imprinting machine with two one-eighth inch diameter pins. One pin fit through the hole in the right end while the other end was held in place by a pin in the notch. As most soldiers never saw dog tags being made on the spot, the function of the notch was somewhat of a mystery. Army folklore had it that the notch was used to wedge a dog tag between the teeth of a dead soldier when he was buried in the field. This did not make particular sense, but there really had to be some use for the notch and this belief satisfied that need.

CHAPTER 4

A Trip to Florida

The next morning, we were ordered to board a Pullman car in a troop train. We ate in the dining car, but the menu was pre-established. All of our troop train travel was in Pullman cars. At night, when you bedded down, you could hear the clickety-clack of the railroad track and the Doppler ding of the crossing signals. Some times the troop train would have to park on a siding until the regularly scheduled trains had cleared the tracks. This was in the days when the railroads were a major and a very fine way of travel.

On the second day of our journey, the troop train was switched onto a siding in Arizona. We ended up sitting alongside of the Casa Grande Ruins. The Casa Grande Ruins are the remains of old adobe southwest Indian dwellings, which are protected by a galvanized iron roof supported by steel posts. This was an interesting coincidence, because there had been an article about the ruins in the Los Angeles Times Travel Section just a few weeks before.

The next day found the troop train sitting in Texas. Sitting on an adjacent siding was another troop train. The other troop train was also loaded with fresh army inductees. Talking with the troops on the other train, we found out that they were coming from the East Coast and headed for the West Coast. This was just the opposite of our troop train journey. No doubt, a fine example of advanced military planning.

On day three of the trip, our train had stopped on a siding in New Orleans. The layover was to be several hours so we were allowed to roam free in downtown New Orleans. We did not wander far as we knew that we would be in really, really big trouble if we missed the train.

MILTON COOK

The final day of the trip east found us at the end of the rails in Miami, Florida. Then it was out of the train, into the Army trucks an off to Miami Beach. Our travel's end was in South Beach at one of the small art deco hotels. Where, after depositing our gear in the hotel "barrack", still "civilians", we were led straggling to a hotel "mess hall" for lunch. Returning from lunch basic training started immediately with the proper storage of our gear and the making of our bunks.

Chapter 5

Basic Training

From various places scattered across the country we had all been transported to Miami Beach to share in basic training. Basic training, while primarily close order drilling and other training sessions to imbue us with the Army way, was also at the same time an initiation and a coming of age rite. Consistent with such rites there were portions of humiliation and degradation attendant to the basic training. The universal belief is that such treatment is required to change the "confused' civilian into a committed soldier. So with this time-honored practice a line of demarcation is scribed which forever separates the soldier from the civilian. A line sharper than a razor's age, a line sharper than a freshly knapped flint, a really really fine line which forever and completely sets apart the fraternity of those who serve and have served in the military from those who never have.

Miami Beach is located on a man-augmented island. Welcome to the Miami Beach Army Air Force Training Center. The barracks were hotels. The mess halls were hotel dining rooms and restaurants. The latrines were the hotel bathrooms and restrooms. The drill fields were the city's golf courses and parks. At the far north end of the island, beyond the upper limit of the city, were located the army stockade or jail, the rifle range and the gas range. At the south end of Miami Beach, just past the last of the South Beach Art Deco Hotels, was the location of the dog track. I'll wager that the Miami Beach Army Air Force Training Center was probably the only training center in World War II with pari-mutuel betting on its premises. That is not to say that betting did not occur on other military environs.

Our particular barrack was a small art deco hotel located at about the middle of South Beach. There were as many steel army cots and barrack bags that could reasonably be placed in each room. Everyone assigned to the room had to share the latrine (i.e. hotel bathroom).

The mess hall was located in the first large hotel at the north end of South Beach. The food and the utensils were regular army. That is to say, the "chow" was prepared in large batches and plopped onto our mess trays. The mess tray is divided into three containment areas by raised strips. The mess tray is undoubtedly the precursor of the divided TV dinner tray. The mess tray is larger than a TV dinner tray, and it is made of stainless steel. The set of utensils was completed with a handle less mug, a knife, a fork and a tablespoon. When we were through eating, the leftovers were dumped into a GI can—galvanized iron and Government Issue. We then placed our utensils in racks for washing. At all times in the mess hall, music played over the P.A. system. I remember Benny Goodman's Band vocalist, Peggy Lee's rendition of *Why Don't You Do Right* playing endlessly.

To turn us inductees into soldiers, we were subjected to indoctrination and to instruction. Our Drill Instructor was Chinese and a Corporal. We were read the Articles of War, which was the Army legal code. We were no longer directly governed by the Constitution of the Republic, even though we had taken an oath to defend the Constitution. The Articles of War were a different code of law, a subset of the Constitution that abrogated some of the rights guaranteed by the Constitution such as the freedom to come and go as we pleased. Also, under the Articles of War, you could be executed for cowardice. We were also instructed in the Army Rules and Regulations and the proper way to wear the uniform.

One of the aspects of wearing the uniform, which in a way embodies the whole military take on life, was the proper and correct wearing of the web belt on our trousers. The belt had a brass tip attached to one end, while the other end of the belt slipped through the back of a lacquer coated "bright" brass buckle. The buckle is held in place on the free end of belt by an over center notched clamp. The belt was adjusted so that the inner edge of the brass tip just touched the leading edge of the brass buckle when it was fitted around our waist. So now, it really irritates me when I see on TV dancers in military uniform portraying soldiers with a foot or so of the web belt order drill while handling the rifles was practiced over and over. We were instructed and learned how to field strip, clean, and reassemble the weapons. We practiced using the shoulder strap to restrain the rifle in all of the field positions. Attention was directed to squeezing off the shot instead of just pulling—jerking—the trigger.

We were issued our side arms, Colt .45 automatic pistols. The necessary accouterments, which came with the pistol, were a leather holster and two spare ammunition clips in a canvas pouch. The web pistol belt, which had already been issued, had a frog-type clasp. The belt was doubled through one half of the clasp to allow adjustment of the length of the belt to fit the waist. Next to the fixed half of the clasp, in the middle of the belt, there was a male half of a snap. The spare ammunition holder was attached to the belt with the matching half of the snap. Eyelets were spaced along the top and bottom edges of the belt. Hooks on the free end of the belt fit into the eyelets to hold it when the size of the belt was adjusted. The field backpack would be attached to the eyelets in the upper edge of the belt. Various items of equipment could be attached to the eyelets in the lower edge of the belt using a double hook arrangement. The prime equipment, which we were to attach to the belt, was the pistol holster on our right side and a pint canteen holder on our left hip.

Of course, we had to learn how to field strip, clean and reassemble our sidearm. As with the rifle, it was imperative to squeeze off instead of pulling the trigger. Over the next two years, I would work the trigger mechanism of my pistol to smooth out the action. This was just in case it was ever required to actually fire the pistol.

The old Sarge discoursed on the proper care and maintenance of the leather pistol holder. He advised us to use saddle soap and not to use shoe polish on the holster. The saddle soap would keep the leather soft and supple, while the shoe polish would dry the leather up. Well, I bought a tin of saddle soap and applied the soap, and I applied the soap to my holster during the entire time that I was in the Army. The leather did stay soft, but it also became thicker and took on a slightly purple cast. Also, the holster never did shine very well, although no one ever objected.

Our training was somewhat directed to living in the field for at least part of the time. The old Sarge recommended that we get ourselves a shaving stick and a shaving brush. These two items were most suited for the field because they were compact and the soap was long lasting. The shaving stick was a round bar of soap, one end of which was attached to a threaded end piece. A closed tube fit over the soap and was screwed down onto the end piece. The shaving stick package resembled a small under arm deodorant package. In use, the shaving stick was held in the fist with the end of the stick flush with the first finger and the thumb. A wet shaving brush was used to work up a—lather on the end of the fist. Anyway, I bought myself a shaving stick and shaving brush, and I used them for years. I had a very slight beard and so shaved infrequently.

Another aspect of living in the field to which we were introduced was the pup tent. Every soldier was to carry in his backpack one half of a pup tent. One tent end pole—in sections—, one piece of cord to stay the end pole, one set of pegs for one site of the tent and the shelter (tent) half. One edge of the shelter half had a row of buttons inter spaced with button—holes that were concealed under a flap. Two shelter—half's were buttoned together and the two-man pup tent was then erected using the two sets of one half of a pup tent. We more or less set up the tents one time in the soft sand of the drill field. This would be the one and only time that we ever would do anything with pup tents. This particular form of pup tent was first use by the Union Army in the American Civil War. At that time when the pup tent was completed from the two kits it was then known as the whole shebang.[2]

One day, we were loaded into trucks and transported up the island to the rifle range. The trip to the rifle range was just to expose us to the various weapons with which we should be familiar, and no scores were to be recorded. The rifles, which we fired, were of World War I vintage. We were provided with either a bolt action .306 Springfield or a bolt action British Enfield rifle. We also fired the Colt .45 automatic pistol and the Thompson .45 caliber submachine gun with a 20 round clip. This would be the only time I would ever fire the "Tommy gun."

Training films were another way of providing information that, it was assumed, we needed to perform our military functions. In addition to the run of the mill training films, the "propaganda" films, the "Why We Fight" series was also screened. Remarkably, I would

---

[2] See Endnote 1 for source such as it is.

tucked in and then wrapped around the belt. In wearing the web belt, the edge of the buckle must be aligned with the edges of the shirt and the fly, and they must align each to the other. To this day, I must align the edges of my shirt, my belt buckle and my fly.

We were also required to learn the General Orders and recite them on command. The General Orders applied specifically to Sentry Duty. One, but facetious, General Order was "When confused or when in doubt, run in circles, scream and shout." After training, I was never in a situation where the General Orders actually applied, nor was I ever after required to recite the General Orders on command.

We were instructed how to and then required to perform calisthenics, including the favorite of the Drill Instructors, the side-straddle hop. We learned and performed all of the close order drill moves. We learned a bunch of marching songs for after all, this was the musical version of World War II. The order when we were allowed to rest was, "Take a break, smoke 'em if you've got 'em."

Our musical repertoire included a number of songs with a cadence for marching. [1] We marched to *Over There*, which was a World War I song. The song was also current from the movie, *Yankee Doodle Dandy* for which James Cagney won the 1942 Oscar for Best Actor. There were two old familiar horse soldier songs, *She Wore a Yellow Ribbon* (cavalry) and *The Caissons Go Rolling Along* (field artillery), which we belted out. *I've Been Working on the Railroad* was another old familiar song to which we marched. There were two songs of French extraction, *Viva La Companie* and *Allouette*, which we also sang as we marched about. Then, there was the silly RAF marching song, *I've Got Sixpence*, (AKA British Flyers Ballad) which goes like this:

---

[1] The lyrics of the songs that we sang and many more that we did not sing were printed, 33 in all, in a thin $3\frac{1}{4}$ X 4 inch booklet sized to fit in to a shirt pocket. A lot of time had been spent selecting the songs, typing the lyrics in a small font (maybe 10 point, but what ever was available on a typewriter), running 2 sided mimeograph copies, cutting, folding, assembling and stapling these 16 page booklets. I mailed my copy of the songbook home and then forgot all about it, but my sister Joy had saved it in her scrapbook.

I've got six pence,
Jolly, jolly six pence
I've got six pence to last me all my life.
I've got tup pence to lend, and tup pence to spend,
And tup pence to send home to my wife, poor wife.
I've got no cares to grie-ee-eve me,
No pretty little girls to dece-ee-eve me,
I'm happy as a king, belie-ee-eve me,
As we go rolling, rolling home.
Rolling home by the light of the silvery mo-oo-n.
Happy is the day when the Airman gets his pay,
As we go rolling, rolling home.

One of the things we could do in the afternoon, when we were free from training, was to go swimming in the ocean. The water was warm but the beach was slightly shoaling. You had to walk far out from the shore to reach a depth that was suitable for swimming. The beach was composed mostly of crushed seashells and decomposed coral. There were a few small blue jellyfish drifting by, which could be easily avoided. In addition to swimming, our other off-duty activities included, going to a movie or wandering around the streets and looking in on the stores and shops. I remember the fresh fruit stands with their bouquet of the tropics.

From South Beach, we were moved up the island to one of the newer and taller hotels. This was to be the start of the next phase of our training. Our new Drill Instructor was an "old" Buck Sergeant. It was my impression that he had been a private for a long time before the start of the Army buildup for the war and was promoted as a reward for his long service.

One of the guiding principles of the military establishment is that everything must be neat and tidy. To this end, we were instructed many times to rip apart our cigarette butts and to let the tobacco and paper fly in the wind. This operation on the cigarette butts was called "field stripping." The old Sarge was always telling us to "field strip our buttses." This amused us all. Rifles were issued, temporarily, so that we might be instructed in all of the aspects that we would ever want to know about the weapon. All of the standard moves in the Manual of Arms were repeated until we could perform them in unison. Close see the "Why We Fight" series again in England as part of the show at the public cinema. The films were shown to the Brits with the explanation that they had been prepared to inform the American troops about the war. We were also subjected to the requisite VD[3] movies. The only thing that I recall about the VD film was the time passing sequence. A GI sets a lit cigarette on a banister rail and enters a room. The door opens, he comes out, straightens his tie, picks up his smoke and flicks off a length of ash. The irony, of course, is that over all in the long run smoking is usually more hazardous than VD. The Army applied substantial resources to the prevention of VD, much like the present war on drugs, and with similar results.

Now we were all ready to be part of an actual unit in the United States Army Air Force.

---

[3] ANA STD, but I will use the vernacular of that "time", as it is pertinent to a later historical point.

Chapter 6

Formation of a Company and Additional Training

New Military Police Companies were being formed. The TO (Table of Organization) established the complement of the company at one hundred men, divided in to four platoons. The TO specified the number of each rank for both commissioned and NCOs—Non-Commissioned Officers—in the company. The Specialist positions such as the Company the Clerk and the Cooks were also enumerated.

A cadre consisting of commissioned officers and some senior NCOs had been mustered as a skeleton on which the new company would be assembled. Some of the commissioned officers were transferred from other Army Units, and some were new graduates from OCS—Officer's Candidate School. The senior NCOs were also transferred in to the company from other Army Units. Some of the NCOs were also promoted in rank when they were transferred into the new company.

The ranks of the newly formed 1228 Military Police Company were completed by, those of us draftees who had just finished basic training. There were some openings in the TO for lower ranked NCOs. These openings in the TO were filled by promotion of some of the newest members of the company. I was to find out much, much later that the reason I was not promoted to even PFC was that I always appeared to be sleepy. However, I was in fact, but unknowingly, squinting because of the bright sunshine.

After several weeks, the 1228 MP Company was split into two sections. One section remained the 1228 MP Company. The other section, which I was in, became the nucleus of the brand new 1257 MP Company. We moved up the island to a different hotel, which was located on the beach. We were now apparently on a fast track, but we were completely unaware of that fact.

The 1257th drilled and trained as a unit unto itself from that point on. One of the significant subjects for training was chemical warfare. This training, in various forms, would continue for years. We had received gas masks as part of our original issue of accouterments. Now we were to find out all about that marvelous device, the gas mask. The gas mask was contained in a canvas bag worn at the side of a strap over the opposite shoulder and a strap around the waist. A canister containing chemicals to absorb and neutralize the chemical war agents—poison gas—was connected to a facemask with an expandable, corrugated hose. There were adjustable straps on the facemask to hold it snug and seal the edges of the mask to the face. There were two glass eye pieces, which provided limited vision. You had to breathe in through the canister and out through an exhaust valve located in the facemask. All in all, a hot and wretched device was the gas mask.

We had the requisite instruction on the effects and odor of the various chemical war agents. In as much as a verbal description of an odor is somewhat limited, it was time to go to the gas range. It is presumed that one sniff is worth a thousand words. The 1257th loaded onto trucks for the journey to the north end of the island and the gas range. At the gas range, a small glass capsule containing each of the chemical agents was ruptured with an explosive device. When the capsule ruptured, a small cloud of agent appeared. We were required to run through the cloud of each agent and sniff to identify its characteristic odor.

There were long intervals between the short times when each of us were able to smell each of the agents. One of the company officers decided that during the intervals would be a good time to practice digging foxholes. So he set us to work digging foxholes in the ground, which was located downwind of the area where the gas capsules were exploded. The ground was all dredge—fill of sand and broken sea-shells. The ground was soft and the foxholes were easily dug. I discovered that a small blister had formed on my forearm. This blister was probably the result of a minute droplet of mustard gas on the ground. Mustard gas is a blistering agent, which also exists as oil under normal conditions. This was to be the only foxhole I was to dig during World War II.

The Army had done its job, we were now imbued with pride and a sense of unity in the 1257th. We were a varied group of young men and teenage boys. Some were married, and some had children, and they had been drafted even as the debate on whether to draft fathers continued. Not all of the company members met the height and intelligence guidelines for Military Police, of which I was informed at my inducting into the Army. We were a diverse band of assorted ethnic origins from hometowns scattered all over the country. This commingling of soldiers from across the map was very typical of the make up of the Army in World War II; a re-melting of the melting of the melting pot.

One of our comrades was a German citizen. He had been a member of the Hitler Youth Organization before his uncle had sponsored his immigration to the U.S. He was required to take the Oath of Allegiance to the U.S. when he was inducted into the Army, despite the fact that, technically, he was an enemy alien. One of the Second Lieutenants had been a Lieutenant on a New York Fire Department. Two of the so-called enlisted men had been Policemen. One wonders at the perverse logic of assigning men to Army jobs so similar to their civilian jobs.

Our appearance as a proficient unit had advanced to the point that we were allowed to march on the city streets. We were issued brown felt campaign hats of pre-war vintage. The wide brim of the hat provided protection from the increasingly hotter sun. We went about the city putting on close order drill shows in the city parks. But all good things must come to an end. There was no graduation ceremony and no passing in review. We turned in the campaign hats, packed our gear into our barracks bags and climbed up in to the back of trucks for the short trip to the train station in Miami. We were off-headed north for parts unknown. You can only head north on a train out of Miami.

Most likely it was a short time before the 1257th moved out from Miami Beach that we were issue our **Soldiers Individual Pay Record.** The pay record was a three-inch wide by five-inch long thin eight-page booklet, which also served as a secondary form of identification. The principal purpose of this document was to provide a means for issuing payment to a GI whose status was casual or straggler, and who therefore was in someway separated from his service record. A casual soldier was either unattached to any unit and thus awaiting assignment or otherwise separated from a specific unit. This separation was usually because the GI was then transient to a newly assigned unit, but in some cases it was detachment for TDY (temporary duty) with some other unit or even a special duty assignment. A strangler was usually a soldier who had become separated from his unit in the confusion of combat and who had temporarily joined up with a unit other that his own. Stapled to the inside of the pay record back cover there was had folded three and a half inch wide by eight inch long immunization register. This register was used to maintain a running history of all vaccinations that had been administered to the GI. The immunization register was, in most cases, the only and real use of the **Soldiers Individual Pay Record.**

PAY RECORD BOOKLET

Chapter 7

Trip to Minnesota

Midmorning on the next day, our Pullman cars were setting on a railroad siding in the Atlanta, Georgia Railroad Station. The Pullman cars had been cut out of the train from Miami, and they were now waiting for another train, which would convey them to our next destination.

We were ordered to dress in our suntans and assemble in company ranks. With "Gideon" flying, we marched off through the streets of Atlanta. The Gideon was a small green flag with the number 1257 denoted on it in gold. Green and gold are the Corps of Military Police colors. The flag was attached to a staff with a spear-point devise at the top. The Gideon is carried by the right front soldier in the first rank at the head of the company.

As it would turn out, we had been headed to the YMCA to go swimming. I suppose that the swimming was intended to be both a treat and a means to keep us occupied during our short layover in Atlanta. The YMCA swimming pool was indoors. It was a good thing that the swimming pool was indoors, because in those days many worlds ago, all swimming at the YMCA was in the nude. While swimming in the nude, one must exercise care when diving into the pool. So there we were, skinny dipping in the line of duty. After our "refreshing plunge", we dried off, put our uniforms on and marched back through the streets to the railroad station, and boarded our Pullman cars.

Soon, in our "new" train, we moving again still heading north. The next day, we stopped in Minneapolis/St. Paul where our cars were switched in to still another train. Shortly after the cars were hooked up the train headed north on the last short leg of our trip to a siding in Camp Ripley, our destination. Camp Ripley is located a short distance north of the town Little Falls. Little Falls is situated about equal-distance between Brainerd to the north and St. Cloud to the south. This is the general area of Minnesota where the mythical lake and town of Lake Wobegon would be located.

Camp Ripley was normally a Minnesota National Guard Camp, but now was being utilized by the Army Air Force as a Military Police training station. Each MP Company at Camp Ripley was a unit unto it self and trained independently. There was a separate encampment for every MP Company that was in training at the Camp. Each campsite consisted of a multi purpose building located on one of the roads that ran through the camp. Situated behind this building were the washing table, the mess tables and a double run of concrete floored tents facing each other across a concrete walkway.

The multi purpose building contained the latrines and the kitchen. The kitchen provided only the basic cooking equipment of a range, an oven and sinks. Only on this one occasion would the 1257th company cooks use the company cooking utensils at the company level. This interval at Camp Ripley was the only time the company cooks ever cooked at the company level. After we arrived at our final destination overseas, they, along with the company cooking equipment, would disappear into the station mess hall "never to be seen again".

The washing table was a large sloping galvanized surface with a raised rounded edge. There were water faucets located on the upper side and drains on the lower side of the washing table. We washed our own underwear and fatigues using GI soap. GI soap came in blocks about three by three by six inches. It was a strong all-purpose soap that was used to clean everything. The soap was a kind of yellow and brown and was semi-translucent. Our suntan uniforms were sent to the base laundry, as they had to be pressed.

The mess tables had attached benches and were located under a canvas awning. The awning provided protection as much from the rain as from the sun. It was now that we started eating out of our mess kits. For the remainder of the war, most of our meals would be eaten out of our mess kits. I estimate that I ate about 2,000 meals out of my mess kit. All of the traditional holiday meals were served onto our mess kits.

The mess kit consisted of the meat can, a knife, a fork and spoon, and a canteen cup. The meat can was incorporated of two separate parts. The bottom part was oval except for a short straight section on the longer dimension. This part of the meat can was one and a half inches deep with rounded sides. The container was six inches wide by eight inches long, with a rolled lip on the upper edge. There was a handle attached to one end of the bottom part with a hinge. The other end of the handle was formed into a tang, which snapped over the lip on the opposite end of the container. The bottom half of the meat can also served as a frying pan in the field. The container was normally made from stainless steel while the handle was galvanized iron. A five-sixteenth inch diameter hole in the end of the handle could be used to hang up the mess kit.

The top half of the meat can was a seven-eighths inch deep oval. When upside down, the top half container just fit into the bottom half. This container had two sections separated by a land. The closed handle of the lower half of the meat can fit into the slot behind the land. A "D" ring was attached to the under side of the land. The "D" ring was attached at the end of the land so that when flipped out, it would stick out past the end of the top half. The top half of the meat can was usually made of aluminum.

The seven-inch long knife, fork and spoon were punched from one-sixteenth inch thick stainless steel. The knife had a molded on black plastic handle. There was a rounded end half-inch wide by one-inch long slot in the end of the handle of each of the utensils. The letters "U.S." were imprinted just before the slots on each of the handles.

The canteen cup had a kidney shaped cross section. The cup slipped over the outside of the canteen. Normally, the canteen and the cup were carried in a canvas pouch, which was attached to the web belt. The canteen held a pint and the cup held about the same. A three section right-angled handle was hinged to the rounded lip on the concave side of the cup. In the closed position, the handle snapped over the outside of the cup. In the open position, the handle was held in place with a sliding latch, which blocked the hinge. If the handle was not latched correctly—disaster! This was a mistake that would not be made more than one time. The handle was made of three-quarters inch wide by one-sixteenth thick metal. There was normally a rounded end, half inch wide by one inch, slot near the end of the third section of the handle. The canteen cup was normally made of aluminum.

To eat a meal, the mess kit had to be set up and held properly. The "D" ring on the upper half of the meat can was placed over the tang at the end of the handle of the lower half of the meat can. The "D" ring slid down the handle until the upper and lower halves of the meat can touched each other. The handle was held against the back of the upper half of the meat can by firmly holding the thumb on the top surface and the first finger on the underside of the lower meat can handle. The remaining fingers were used to support the bottom of the upper half of the meat can.

With your "silverware" tucked into a pocket, your meat can assemblage in one hand and your locked canteen cup in the other, you entered the chow line. The cooks and KPs plopped the food into your meat can and filled your canteen cup. Then it was on to the mess tables to enjoy the feast.

After completing the repast, it was time to clean the mess kit. Cleaning the mess kit was accomplished via a triad of GI cans. Can number one was for leftovers or any other food unfit for consumption. Said food was removed from the meat can using a tethered large stiff bristle brush. The next GI can was filled with hot—to the point of boiling—soapy water. The last GI can contained hot—to the point of boiling—clean water. I had never thought about it before, but there must have been standard heaters[4] for the water-filled GI cans. I encountered this triplex of GI cans in many locations and their conditions were always the same.

To wash the mess kit, the top half of the meat can was slid down the handle of the bottom half on the "D" ring. The top half rested on the bottom half of the meat can at an angle. The slots in the handles of the eating utensils and in the handle of the canteen were used to hang them from the handle of the lower meat can. Then, it was dip and swirl in the each of hot water filled GI cans. The mess kit would dry almost immediately. At this point, I encountered a significant problem, as my canteen cup did not have a slot in its handle. I had to wash the canteen cup with one hand—getting dangerously close to the very hot water. At the same time, I had to dip and wash the rest of

---

[4] White gas (see end note 2) fired immersion heaters were used, so my brother Allen informed me. He had worked on these water heaters when he was in the U.S. Army.

my mess kit. It was my conjecture that my canteen and canteen cup were left over World War I equipment. Later, once over seas, I took care of the problem by cutting a slot in the handle of the canteen cup. As it turned out, the handle was made of nickel-plated brass, and an egg-beater drill and a file did the trick.

We were quartered in standard U.S. Army square canvas tents with pyramidal tops. The tents were supported on a fixed wood frame. The tent frame was attached to a concrete pad. The pad was connected by a short walk to the central concrete walkway located between the two rows of facing tents. The tent walls were six feet tall with roll-up flaps, over screen, to provide ventilation. There was also an installed doorway with a wooden door. Hanging bare light bulbs provided illumination. However, lighting was not needed much during the summer at the latitude of Camp Ripley. The days were long and the nights were short. The light did come in handy during the thunderstorms.

Our bunks were standard army folding canvas cots with a standard thin pad U.S. Army mattress. Over each cot, there was a mosquito bar. "What the heck is a mosquito bar," I wondered when I first heard the term. As it turned out, it was a mosquito barrier. The bar consisted of a "T" framework attached to each end of the cot over which was draped a rectangular mosquito net.

When the day's activities were completed and on weekends, unless there were assigned duties, you could get a pass out of the camp. There was a regular bus service from Camp Ripley to Little Falls. The town is located on the upper reach of the Mississippi River, which was not mighty at this point. There were little falls in the river, as the name of the town implies. The side of the tank on the town water tower proudly proclaimed "Little Falls, Home of Charles Lindbergh." There was one main street with motion picture theaters, restaurants and stores in Little Falls. When not involved with these activities, we would just walk around the town seeing what we could see.

When you see a film with multiple columns of soldiers marching down a wide boulevard, and they turn the corner smoothly, it's because they had practiced. You have to be prepared for a Victory Parade. As part of our advanced close-order drill, we practiced making the multiple-column turns. Notwithstanding, that at the company level, the columns are much shorter than they are wide. However, the turning conditions are the same where the pivot man is marching in place and the outermost column is almost running. We also practiced advanced close-order drill with fancy manual of arms until we were really good. But, never would we apply these drills in actual practice.

We also practiced drills unique to military police activities. Drills of point control were performed. Point control is similar to being a traffic cop, except it more frequently involves the direction of troop movements. We practiced the hand signals used to direct the Army units and vehicles to which way to proceed. There were only a couple of occasions where I performed an activity somewhat resembling point control.

We had been issued nightsticks as part of our military police equipment. The nightstick in current parlance is referred to as a "baton." It was, in fact, just a nicely turned section of broomstick. The nightstick had a multi grooved roundel hand grip, was stained brown and had a closed loop thong through a hole below the handle. The thong loop was used to attach the nightstick to the belt. In use, the thong was wrapped around the hand to prevent the nightstick from being pulled from the hand. We learned to whack on the arms and shoulders, but not on the head with the stick. The stick could be held horizontally as a block and could be used to pike. We went through all of the drills, including retaining the nightstick when someone tried to pull it out of your hand. I carried a nightstick on duty throughout the war and never used it hit anyone or anything.

In the company armory, there were a number of twelve-gage automatic shotguns. These U.S. Army riot guns[5] were fitted out with a standard rifle shoulder strap, bosses on which a bayonet could be installed and a perforated metal guard over the gun barrel. I suppose the barrel guard was to prevent burning yourself during sustained fire. The projectile to be used in the gun was a big game load consisting of sixteen thirty-two-caliber balls contained in paraffin wax. We practiced advancing on the "mob" as a diagonal

---

[5] Probably these shotguns were left over from World War I when they were called trench guns.

line with the shotguns held at waist height. A line of military police advancing with shotguns, which bore bayonets, should give any mob pause. But, alas, it was not to be; I never saw the shotguns again after this drill.

Our primary weapon was our sidearm, the Colt .45 automatic pistol. Camp Ripley did not have a regulation pistol range. That being the case, we could not qualify for proficiency with the Colt .45. We did spend some time firing the pistol at targets. While we could not establish proficiency with the .45 automatic, we did develop some skill at hitting the target.

Camp Ripley did have a regulation rifle range, so even if proficiency with the pistol was out, we could become qualified with the rifle. (It is possible that the MPs were intended to provide an infantry function for the Army Air Force should the need arise.) The major feature of the rifle range was two parallel, high, earthen embankments. The targets, the target handlers and the scorekeepers were behind one embankment. The novice riflemen and the range safety officer were behind the other embankment. The major function of the range safety officer was to make sure that the rifleman kept his loaded weapon pointed towards the opposite embankment at all times.

The shoulder strap was used to hold the rifle snug against the shoulder. Firing was performed from the standing, the kneeling and the prone positions. The Garand M1 was a semi-automatic rifle having a rapid fire and a some—what light recoil. However, the thirty-caliber M1 rifle did have one small human factors problem. The Garand was loaded by using the thumb to push an eight round ammunition clip into the cavity behind the breach of the rifle. Just as soon as the ammunition clip was completely inserted, the breach block slammed forward very rapidly. Considerable agility was required to remove the thumb from the path of the closing breach block. Having your thumb smashed provided a great incentive not to dawdle. Reloading the Garand while lying on the ground was especially irksome.

The targets were supported between two posts. The targets were counter weighed to ease their movement up and down behind the embankment. After the rifleman had fired a shot, the target was lowered. A large circular marker, on a short round stick, was used to mark the bullet hole. The target was then raised up to show just how far the shot was from the point aimed for, the "bulls eye." The target was then lowered, scored, patched and sent up for the next shot. A small flag on a long pole was waved, for all-of-the world to see, when a wild shot had completely missed the target. This flag, which was waved in front of the target, was known as "Maggie's drawers."

There are three levels of qualification for which shooting badges are awarded: Expert, Sharp Shooter and Marksman. A bar with the name of each weapon for which you are qualified is attached with links to the bottom of the shooting badge. I qualified with the rifle as a Marksman. However, badges were not available at Camp Ripley, or for that matter, at any other place where we were during the war.

On one weekend in early August, I performed my first function as a Military Policeman. It was now time to officially wear my MP bassard. The MP bassard is worn on the upper left arm. The band was a black wool three and three quarter inch wide cloth strip. Two and a half-inch tall block letters "MP" in white wool cloth were sewn onto the black strip. The back of the arm bank was lined with khaki cotton cloth. Two buttons were sewn on the back of one end of the bassard. Two rows of buttonholes pierced the other end of the bank. The multiple buttonhole arrangement made the armband adjustable to fit on sleeves from shirts to overcoats and a range of arm sizes, except when there were no buttonholes and safety pins were used.

We loaded into a truck for a jaunt down to St. Cloud. We were only armed with our nightsticks. We did not have ammunition for our side arms, so they would be useless anyway. We were assigned a beat to walk in pairs. As MPs, we would always patrol in pairs. Our instructions were not to enter any bars. Except, that is, in the case of a disturbance, when we were asked to come into the bar and take care of the situation. We could, however, go into a restaurant to relieve ourselves. It was very tedious just walking back and forth through a section of downtown. I dozed off momentarily while walking. I had never dozed while walking like that before, and I never did it again. But, it was a weird experience to nap while walking.

Well, again there was no parade and no passing in review. However, shortly before departing from Camp Ripley, we did assemble for the requisite company picture.

1257 MILITARY POLICE COMPANY (AVN)
UNITER STATES ARMY AIRFORCE
CAMP RIPLEY, MINNESOTA
AUGUST 1943

My platoon marched to the armory where, on the drill floor, we packed all of the company equipment back in to its shipping containers. Then on returning to the company area we loaded our gear and ourselves in to the back of Army trucks for the short drive to the Camp railroad siding. Then it was load back on a train in Pullman cars for a trip south and east heading for the inevitable sea voyage.

PART
III
AN ATYPICAL ASSIGNMENT

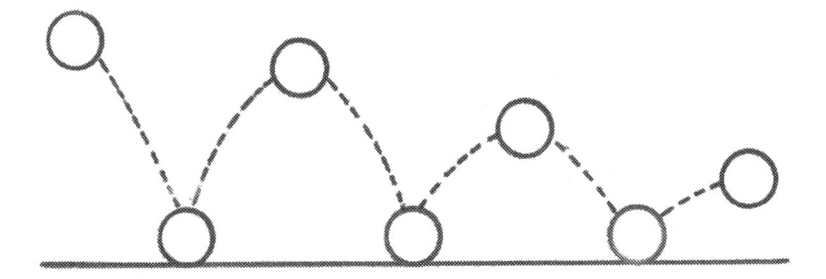

PART III
INSTITUTIONAL ASSOCIATION

CHAPTER 8

Camp Miles Standish, Massachusetts

The next morning, the train had stopped in the Chicago railroad yards. Here the Pullman cars were switched from our train of arrival to another train for our departure. Chicago, the same as London[6] and some of the other older metropolis, did not allow railroad trains to pass through the center of the city. There is not even a central railroad station in the city. The usual way to pass through Chicago on a train was to change trains. You would arrive in one station, detrain, take a bus to a different station, retrain, and then depart from the city. That is unless the car in which you were a passenger was switched, as ours was, from one train to another train in the railroad yards. The train to which your car was switched then took the freight train route around the city.

The 1257th MP Company was a small contingent and therefore, did not pause in Attleboro. Instead, we continued straight away to Camp Miles Standish. We were quartered in a long, single-story wooden barrack. Our particular barrack building was the second over from the building, which was the central latrine. If events transpired that you must pay a visit to the latrine after taps, you had to first get dressed. Then, exiting through the back door of the barrack, you needed to scurry along the dirt path through the dark of night to the latrine.

---

[6] The cities were already there when the railroads came to town and the city fathers did not want to be disturbed by those noisy and foul stench trains. This was in the face of the fact that horses were the mainstay of transportation and that livestock was also driven through the streets.

The next morning, we were sent up the road to the administrative offices to spend some time with the clerical staff. We had to execute paperwork related to our pending, protracted non-residence in the United States of America. I verified my mother as beneficiary of my $10,000 National Service life insurance policy. The most significant document, which I completed, was a power of attorney to my mother.

The road to the administrative offices was past the PX. I stopped at the PX as I made my way back to our barrack. The PA system was playing the song *June Is Busting Out All Over*, over and over endlessly. This song was completely new to me; as I had never heard it before, isolated as we had been from the main stream while being subjected to training. During our training, we did not have radios available. There was a scarcity of new papers and magazines, which I wouldn't have read during our training in any case. The major source for information of what was occurring in the nation and in the world was the newsreels at the movies in town.

Most of the lower ranking "enlisted" members of the company spent the better part of the day loading the company equipment into a railroad boxcar. The company equipment had been packed into five heavy boxes. The boxes were constructed of thick pine, and skids were attached under the shorter sides of each box. Each box was a different size, shape and weight. One box was long, low and narrow. The biggest box was almost square on all of its sides. The shapes and the sizes of the three remaining boxes was enveloped by those the first two boxes. We slid these five boxes into and out of the railroad car time after time. It was like a Chinese puzzle, which we could not solve. Try as we might, we could not find a way to fit all of the boxes on the floor of the boxcar. The solution was to rest one end of the long box on the shortest box.

Each of us in the company was issued twenty-one rounds of .45 caliber ammunition. Seven rounds each went into the clip in our sidearm and into the two spare clips. This was our graduation. We were now no longer in training, but were full-fledged soldiers. I was now a teenager with a loaded gun. This was a somewhat strange state—being armed. It was as though we could not be trusted with live ammunition in the USA, but now that we were leaving the Zone of the Interior we could be trusted. In any case, as soon as we cleared the port of debarkation, we would be in a war zone and, therefore, in harm's way. We would carry our loaded side arms until reclaimed by the Army prior to returning from Europe.

The most probable destinations when shipping overseas from the East Coast of the United States were Sicily, North Africa and England. Our probable destination was clarified by two events: first, we were required to return our suntan uniforms to the Army Supply Depot; and second, we were each issued a pair of black Arctic overshoes. Taken together, these two events ostensibly eliminated Sicily and North Africa as our final destination. However, it could have been the Army trying to confuse the Nazi spies as to just what was our mission.

Chapter 9

Across the Atlantic

The fate of war shuffles the deck and deals everyone a hand, and you have no choice but to play the hand you were dealt. Some will spend the entire war a short commute from home and some will never ever come home. It's a strange game to play, for we do not know what cards we hold, and the first rule of this game is that all of the other rules are subject to change without notice or reason. As fate would have it, some hold better hands than others do. However, you could have a re-deal if, when the call went out for volunteers, you accepted.

As it would turn out, our hands contained some wild cards, but at the time I did not think that this could be the case. It was now 165 days after my induction into the Army, and there had been no overnight passes or furlough.

As we entered the gangway, through the side of the ship, to board His Majesty's Ship Queen Elizabeth, the American Red Cross handed each of us an OD drawstring bag filled with goodies. As I recall, the bag contained things like a shaving kit, soap, playing cards, cigarettes, candy, etc. This was a wonderful gesture, and a fine gift from the American Red Cross and the American people. But I thought to myself, anyone who did not have the foresight and good sense to take such things with him when embarking on this journey would probably not survive the war.

We climbed up many stairs and through quite a few decks to reach our compartment. The compartment was cavernous and dark. The only illumination was provided by a few always-on ruby red lights mounted near to the deck. This chamber had been a first-class cocktail lounge in those jolly days before this Queen had gone to war.

The compartment was filled with ladders of bunks stacked five and six high. The beds were typical nautical construction with a grommet edged canvas hammock that was rope laced inside of a rectangular pipe frame. The frames were hinged to stanchions on the inner side and supported with chains from the stanchions to the outer side of the bunks.

We were some of the first troops to board the ship and so had a lot of time on our hands until the last soldier came up the gangway and the ship got underway. So, we availed ourselves of this waiting time to man the rail to see what was happening in this port of New York. As the day advanced, more and more troops assembled on the pier and, in single-file, wormed their way through the boarding door to their berths located somewhere in the ship's maze of decks and compartments. And then, on the other side of the pier, there was a strange sight. For there she rested, the rusty hulk of the French liner Normandie, which had been renamed the Lafayette as she was to be used as a troop ship. During the process of conversion in February 1942, she caught fire and capsized. The super structure had been removed, and the hull was in the process of being re-floated and righted. As it would turn out, the rehabilitation would cost too much, and she would be scrapped out in 1946.

Some time during the night the ship had gotten under way and slipped out of the United States, or in the vernacular of the U.S. Army the Zone of the Interior or the Z of I or the ZI. Thus we were now in a war zone and so would be until victory was declared. On the morning of the twenty-first, the ship was alone and beyond the sight of land. We were still close enough to the U.S.A. that blimps could be used to provide our air cover. On the subsequent clear days, an aircraft was always seen somewhere in the sky.

This was to be no pleasure cruise as the 1257th was to be the police force for this trip. We had to be on duty while the rest of the GIs on board were just going along for the ride. Some may have had KP; I don't know. My post was by a PX on the promenade deck immediately aft of our quarters. I spend four hours every afternoon observing the sky, the sea and the fellow GIs playing cards and reading. One of the booklets which they may have been reading was "A Short Guide to Great Britain." This was a four by five-inch thin pamphlet providing orientation on England and the British. My guesstamate[7] is that the ship was cruising at twenty-five knots. The ship was proceeding on a winding course across the ocean. The ship's speed and the sinuous track combined to make it virtually impossible for a Nazi submarine to score a torpedo hit.

There was no wind on the sea, not even a breeze. The wind had been dead a long time as the sea was smooth. Such a calm is a very unusual sea condition at anytime, but it is particularly strange for the North Atlantic. Looking over the fantail, there was the ship's wake wandering aimlessly across a sea of glass to the edge of the world.

On the morning of the twenty-seventh, the sky was overcast, and the sea was showing some activity. To the south of our course, we were passing a rocky end of the land wrestling with the sea. This land's end was a brilliant emerald green, so that I knew that this must be a northernmost tip of Ireland. Later that morning, we were in the Firth of Clyde heading for port.

The Queen Elizabeth moored in Port Glasgow, Scotland, and soon we disembarked and immediately boarded a railroad passenger car. It was a British long distance coach with a number of compartments on one side and an aisle on the other side. While the train was being loaded with additional troops, young boys were along side of the tracks wheedling cigarettes. I was shocked to see how young these boys were.

There was a billboard, along the side of the railroad tracks, advertising Brille Cream. There was a touch of home. I had used Brille Cream before, and would use it again after the war. This was a time when fashion required men to use pomade on their hair to make it shiny and to hold it in place. But, then we didn't talk about what we used on our hair we just used something. Hell, we were men.

Finely the train, that we had all loaded in to, was completely filled with other GIs, it was off rolling down the tracks. Running quickly through Glasgow and then hurrying southward from Scotland in to England we watched the "different" British countryside and cities slide by.

---

[7] "Guesstamate" is a technical engineering term, which means carefully calculated using assumed reasonable but unverifiable data.

Chapter 10

The Trip to Stone

We were now in the ETO, the European Theater of Operations. Why was it a theater? Our train stopped in Stone, Staffordshire, and we detrained. Stone is located in the English midlands on a line from Manchester to the north and Birmingham to the south, just a little closer to Birmingham. We loaded onto Army trucks for the about two-mile trip to Yarnfield. Yarnfield proper was a small community of a few farmhouses and one mercantile building housing a small country store and a pub. The pub Labor In Vain Inn[8] was off limits to U.S. Army personnel. (*Labor in Vain* would be a good name for a rock group.)

The headquarters and some of the other units of the then Eighth Air Force Replacement Depot—"Repple Depple" in Army-speak—were located in the area of Yarnfield. This particular Repple Depple complex consisted of three billeting compounds with some shared and some duplicated services. These units had been constructed, but never used, as housing for women who were to be employed in one of the munitions factories in the area. These housing units were for the most part named for the past successful British Admirals, followed by the appellation "Hall." The headquarters offices for this and all of the other Repple Depple stations in England and later in France were located at Duncan Hall.[9]

---

[8] The requisite artwork mounted on the face of the pub, in this case, portrayed an English matron scrubbing the back of a young bare black boy in a small tub.

[9] See endnote 3 for an essay on the "Disposition of the Repple Depple."

THE LABOR IN VAIN INN

> This is a very bucolic scene, but obviously posed. There was way too much traffic on the road pasts the front of the inn to allow dairy cows to roam around on the loose.

A short way beyond Duncan Hall, just before the **Labor In Vain Inn**, a right turn into a side street was the way to the other two "Halls" in this station. The concrete paved street passed through the main, and only, gate of Howard Hall, continued on past Howard Hall proper, to the right, and meandered on a short way passing grassy fields to Beatty Hall.

Beatty Hall was the final destination of our short trip from Stone. We alighted from the trucks, the rest of our baggage to follow, and were quartered in one of the housing units. These housing units were barracks by definition only. The configuration of all of the barracks, at all of the stations, of the Repple Depple was virtually the same. The barracks were "H" shaped buildings with a fat cross bar, which extended out from the legs equally on both sides. From the Common's Room, in the cross bar between the legs, hallways stretched down the center of each leg. Small "two person" rooms, with doors, lined each side of the hallways. The extensions of the cross bar beyond the legs were mirror images. In each extension, there was a utility room, with storage for cleaning equipment and supplies, toilet paper[10], and a very large water heater. The greater portion of the cross bar extensions contained the latrines. There was the typical row of bathroom sinks and mirrors along one wall. Unlike the usual GI latrine with its rows of side-by-side commodes and one communal shower, here there were a row of commode stalls and a row of shower stalls.

In a surprising announcement, we were informed that this Repple Depple was to be our assigned billet. We were now a part of a permanent party and not just "casuals" passing through. The 1257th MP Company (Avn) was now part of the 93rd Sta Comp Sq (Sp) AAF Sta 594, APO 635. To translate, it was the 93rd Station Complement Squadron (Special) Army Air Force Station 594, Army Post Office 635. The 93rd Sta Comp was the headquarters unit of the Eighth Air Force Replacement Depot. The other stations of the Repple Depple at the other locations had their own station complement.

The first assigned duties for the 1257th was to man the gates of Howard and Duncan Halls. The Office of the Provost Marshal and the Stockade were a large guardhouse situated at the Howard Hall gate. The guardhouse was the police headquarters for the 1257th MP Company. There were two large swinging gates, which were closed across the street and latched at night to stop any vehicular traffic. A small gate, used for foot traffic, next to the guardhouse would also sometimes be closed at night. There was a pass window at the gate, which provided direct communication from the guard to the inside of the guardhouse. There were always guards and other personnel on duty at this guardhouse and gate. Late at night, during cold and rain, the small gate was closed and manned from inside of the guardhouse, as there was no shelter outside by the gate.

Duncan Hall had two gates facing on to Yarnfield Road. During the day, the west gate was the entrance gate and the east gate was the exit gate. There was a small gatehouse on the west-side of the entrance gate, which provided some shelter from the wind and rain but not from the cold. The two swinging gates were closed and

---

[10] The toilet paper was English wartime toilet paper. It was rough and gray with strange spots including small flecks of wood.

locked at night. There was no one on duty at the closed west gate during the night. The east gate, the exit gate during the day, became the entrance and exit gate at night. On the east-side of this gate, there was a gatehouse and a small gateway for foot traffic. The two swinging main gates would remain open at night until the last MP Patrol vehicle had returned to the station. This gate was manned at all times and this gatehouse also was not heated.

STATION PERIPHEY FENCE

A typical section of the "high security" barrier was just the thing to keep cows from wandering to the station. Szucs and Cook hanging on.

Gate duty was a lot of standing and a lot of waiting, which was occasionally interrupted to check documents when vehicles were logged in to and out of the station. There was a rush out through the gate at night when many GIs would have a pass and go to the local towns. When passing through the gate staff cars and MP vehicles did not generally stop. A series of swards, which had once been farms, extending from Duncan Hall at one end to Beatty Hall at the other end, were part of the station. The two units of the station, Duncan

Hall on one side of Yarnfield Road and the connected Howard and Beatty Halls on the other, were each enclosed by fences. These fences consisted of concrete posts and six bare wires. The four-foot tall concrete posts were pierced through with six unevenly spaced of horizontal holes that were closer together near the ground. One-eighth diameter galvanized iron wire had been threaded through the holes and pulled taut to the point where the wires were rigid as rods. The wires and the MPs on the gates provided all of the control necessary to secure the periphery of the station. No one apparently wanted to make an unauthorized exit from or entry into the station.

Well, the Army issues, and the Army takes back, because whatever the Army issues still belongs to the Army. The Army took back the black arctic overshoes, which we had been issued prior to embarkation. I thought that the overshoes were reclaimed because we would not be providing security, in the mud, at some Air Force Base in East Anglia, and therefore, we would not need arctic overshoes. There were, however, several occasions when I had a real need for a pair of arctic overshoes.

Thursdays were gas mask day. The time had long passed when the English population had taken their gas mask with them everywhere they went. But on Thursdays, we had to strap on and carry our gas mask around with us all day. It was a field day, however, for the chemical warfare section, the "stinkers," who drove around the post in a jeep randomly lobbing CS-2 tear gas grenades. In the event that you were in an area where a grenade was tossed, needless to say, you would quickly put on your gas mask. Gas Mask Thursday did not last for any length of time. The gas mask drill was very disruptive to the normal Repple Depple function of processing the ground crews and aircrews headed for the Air Force Bases and the aircrew men who were being sent back home to the States. At this time, any Eighth Air Force bomber crewmember, that had completed his twenty-five missions over Europe, was returned to the United States.

Another unusual "training" exercise was to introduce us to the then current field rations. Each C-Ration meal came in two round cans. The cans were about three inches by three inches and had winding strip openers. One can contained some sort of hard crackers and a small metal container. The small container held powdered coffee in the breakfast meal and lemonade mix for another meal. The second breakfast can was filled with some type of egg and meat concoction. The second can for the other meals contained some kind of hash or stew. We ate our sample of C-Rations, and these were haply the last C-Rations we would ever see or have to eat. It was just as well, because as I remember, these rations were not very tasty. We were also shown a D-Ration. The D-Ration was a very hard quarter pound chocolate bar, in a waterproof container. Also, this would be the last D-Ration that I would ever encounter. We would however, several times in the future, be issued K-Rations for actual meals.

CHAPTER 11

A Trip North to Chorley

We had been quartered in Beatty Hall for a few weeks when the men in my platoon were ordered to pack our gear. We were being placed on temporary duty to provide MP support at another station of the Repple Depple. Our new temporary station was located at Chorley, Lancashire. Chorley is northwest of Manchester and Northeast of Liverpool with the three municipalities about equal distance from each other. This station was always referred to as Chorley although its actual name was Washington Hall. Generally, if there was a sizeable town near to a military establishment GIs would refer to it by the name of the town, as this identifies its location. Thus, the headquarters complex at Yarnfield was referred to as Stone throughout the Army Air Force in England. Similarly, I learned to say Los Angeles whenever asked where I was from because only an identifiable general area of the country was of interest.

Our "new" station was located on the outskirts of Chorley, just before Euxton, on the road to the west. Euxton was a small outlying group of homes, a store or two and several pubs on the road a short way past the station. The gates to the station and the gatehouse were located on the road a few hundred yards from the main body of the station. The gatehouse had windows all around from waist height up. A flashlight was required to make log entries during the darkness of the night. Late at night after "taps", the stillness was interrupted by the sounds of bells ringing across the fields. A carillon on the unseen church marked the quarter hours with progressive chimes and announced the hours with pealing bells. In the middle of the night and the darkness of the blackout, the ringing of the church bells across the meadow was the only sign of life outside of the gatehouse.

Except for one very strange occurrence on a very dark night when an old crone came up to the gatehouse. One of the MPs who were on duty that pitch black night was Little Joe, a GI of Sicilian heritage. And as it turned out the other MP who was on duty had to restrain Little Joe, because he was convinced that this woman was a witch and so was going to shoot her.

It was at Chorley that our helmet liners were painted white and embellished with "MP" in black on the front. Whiting, which was used extensively by the British military, was provided to us with instructions that it was to be applied to our leggings and our web belts. The whiting, which came in small wide-mouth jars, was similar to white shoe polish. It was a water-based emulsion that produced a bright white coating that did not chalk off. All of the U.S. Army MPs in England were to be similarly arrayed. The MPs in London also had a white shoulder strap in a Samuel Brown arrangement. White equipment was required on the MPs in order that they might be easily identified at in the blacked out night.

At Chorley, the Station Commander wanted a retreat ceremony, so a retreat ceremony was performed every day. While marshal music was played over the PA system, three of us in full MP regalia would march up to the station flag-pole. When the PA system sounded the retreat bugle call, one MP would lower the American flag. The second MP would grasp the end on the flag, and the third MP would unsnap it from the line. The two MPs holding the ends of the flag would then fold it ceremoniously. The other MP would grasp opposite forearms to form a shelf. The folded flag would be placed on the folded arms with one point out. The three MPs, one on either side of the flag bearer, would then march to the guardhouse and deposit the American flag there for the night.

On the way into town, there was a park on the north side of the road at the edge of Chorley. This park was a small rounded bottom valley, all green with grass and trees. Pathways wandered through the park and a small stream made its way along the bottom of the valley. On the other side of the road, across from the park, there were a series of gas streetlights—maybe the last of their kind. What was even stranger still about these streetlights was that they had not been on for over four years and yet the pilot lights were still lit, casting a little flickering light on the ground below. It may have been that the pilot lights were connected to the main gas line, and there was no way to turn them off in the light fixture, or maybe someone had just forgotten to turn them off.

At about the edge of the park the road sloped on down as it went in to Chorley. The town was predominantly red brick with an overlaying gray coating, as was most of Great Britain—the accumulation of over a hundred years of coal smoke, ten years of "the slump"[11] and four years of war. A major attraction in town for the GIs was, of course, the pubs, of which there were several. The other attractions for the GIs were the women, of which there were many.

As MPs, we almost always worked in pairs. There were two men with assigned duty at the gatehouse twenty-four hours a day. At night, foot patrols in Chorley and Euxton and on the road past the station were always in pairs. The foot patrols started when the men were released from the station with a pass. The pubs closed at ten o'clock, and then it was time for the GIs to start back to the station, and there we were bringing up the rear as the soldiers drifted back to the station. We would urge them to keep walking, as we could not go off duty until the last GI was out of the towns and back to the station. We always called this activity "herding the sheep".

Foot patrol was in reality a very routine activity with a lot of time to kill. There were almost never any problems, and when problems did occur, they did not amount to much. The dearth of unsociable behavior was no doubt due to the short time that the pub's were open, and the fact that only beer was available to drink. These were in any case well-trained and well-disciplined soldiers of the United States Army Air Force who were not prone to getting into trouble.

There were only a few pubs in Chorley, and reasonably we could only show our faces in them from time to time.[12] There was a roller skating rink at which we would spend some time talking to anyone we could and watching the skaters. We could get a British wartime Coca-Cola in a traditional coke bottle at the skating rink. To me, it seemed that the flavor was less hearty than the coke flavor that

---

[11] The British economic down turn was not as deep, nor as prolonged as the Great American Depression.

[12] It was in Chorley that I first discovered, or someone told me, that closing my eyes when going from a lighted area into the darkness of the blackout would speed the adjustment of my night vision.

I was familiar with. I thought at the time that maybe the coke had been made weaker to extend the supply of coke syrup and to use less sugar. (This may not have been the case as I have subsequently been apprised that that at the time the British favored less sweet soft drinks as a mater of taste.)

While staying in range of the pubs, we would wander around and explore the dark and almost deserted streets. The silence of the night might be interrupted by the sound of a British soldier's hobnails grating and screaking across the paving stones. We would visit the dark and empty railroad station and stroll down the empty platform. And always, from deep inside the station, we would hear the stationmaster playing the spoons and singing loudly to himself. We were not the only one's who were hard at it killing time.

A treat, some mornings at breakfast we would be served fried fresh eggs. I thought that these eggs had been in cold storage for a long time based on my experience working for an egg farmer while I was attending high school. The fried eggs that we were served at Chorley were unique because all of the other eggs that I was ever served in an ETO mess hall were in the scrambled form and made from the powdered form. In the mess hall, the de-powdered eggs would appear in large pans about two inches deep by a foot wide and a foot and a half-foot long. It was always my impression, based on the form of the scrambled eggs that they had been prepared by baking. Other than the fact that these scrambled eggs were food, they had no other redeeming attribute

As an accompaniment to the eggs, we would frequently be served sliced "Spam." This was not the Spam that we knew when at home in the U.S.A. This meat had a similar appearance to "Spam" but it was just pork loaf, and it came up short on the flavor. The pork loaf came from long square cans that were about four by four by eighteen inches on the sides. Pork loaf was the meat of choice at many other meals besides breakfast.

When we had first arrived in the ETO, there was a running dialogue in the *Stars and Stripes*[13] about Spam and Brussels sprouts.

---

[13] The *Stars and Stripes* U.S. Army newspaper was our main source of news. The twelve by eighteen inch London edition daily paper was very much like a university daily paper. The paper provided coverage of world

THAT'S THE WAY THE BALL BOUNCES

Rations had been in short supply due to the supply problems in the early days of the influx of American forces into England. During this time, Spam and Brussels sprouts were major staples and had become a subject of disdain. I, myself, did not encounter very many Brussels sprouts in my meals, because by the time we had arrived in England, the logistics of food supply had improved. However, before we departed from England and the major part of the troops had returned to the United States, we would be on short rations due to the deterioration of the food supply logistics.

While we were stationed at Chorley, the enlisted men and the causal commissioned officers shared the same mess hall. There was a separate mess and a designated section of the mess hall for each group. The enlisted men ate out of their mess kits. The causal commissioned officers had their food served on mess trays. The permanent party commissioned officers dined in their own private mess with their food served on china on a white tablecloth.

Chorley [14] was the Repple Depple station that processed the transfer of Americans, who had been serving in the British and Canadian forces, into the American Army Air Force. Every day in the mess hall, I would see the blue uniforms adorned with British and Canadian Airman wings and decorations, on Americans serving in

---

news from the wire service, but the main emphasis was on the progress of the war in the ETO by the local Army reporters. After all, we could not go home until this war was finished and going home was what was important. News of a more general nature, humor and feature stories was provided in the British edition of *Yank the Army Weekly*, 3d. Yank was a ten by thirteen and a half inch twenty-four page magazine similar to a Sunday supplement.

[14] During the time while I was stationed at Chorley, I saw two unusual, to me at least, meteorological phenomena. One event, occurred very early in the morning, when there was a low wispy fog drifting across the grass-covered field. One of the locals told me that what I was seeing was a Scotch mist. I have never observed a Scotch mist again. The other event occurred at night and was even stranger. There was a small amount of fog, and the full moon was shining low in the sky. On the opposite horizon from the moon, there was a gray moon bow with the same shape as a rainbow.

the RAF and RCAF who were now being transferred into the Army Air Corps. These men, both the commissioned and non-commissioned officers, were for the most part Americans who had signed up with the RAF or the RCAF before the United States had gone to war. The commissioned officers were exchanging their blue uniforms for OD service jackets and pink whipcord trousers. Also, British service men had with at least one parent who was an U.S. citizen were eligible to and did some times transfer into the U.S. Army, even if they had never been to the United States.

Apparently, some of the RAF and RCAF non-commissioned officers were receiving direct commissions as officers when they transferred into the Army Air Corps. I think that this was so because of the following occurrence. Two of us were assigned foot patrol duty in the area of Euxton. While making our rounds, we encountered a group of officers and exchanged salutes. At this point, each of the officers tried to present a ten-shilling note to each of us. We refused the notes because we were on duty, and therefore for some reason we could not accept the money. But to tell the truth, we did not have any idea of what had occurred. We thought that maybe they were trying to bribe us for something, but we knew not what. All of our time in the Army had been out of the main stream, and so we were not aware of an old Army custom. In fact, I only understood what had occurred when I read about the custom many years later. What had occurred was that these officers had just been commissioned, and the two of us were the first "EMs" to give them a salute. The old Army custom was that a newly commissioned officer must give a dollar bill to the first EM with whom they exchanged salutes. So we missed out on a few pounds because of our unawareness of an old Army tradition.

Chapter 12

Back South to Yarnfield

When our detachment returned to Yarnfield from Chorley, we were surprised that the 1257th had been moved from Beatty Hall to Howard Hall. This move was a good deal as the barrack in which we were now quartered was the first one of a double row and the one closest to all of the station facilities. I was assigned to an outside room that was adjacent to the cross bar section of the barrack. These rooms were small, about six by ten feet. Usually, there was a double bunk bed set up against each of the six-foot wide end walls. This was the case for all of the casual "EMs" who passed through the Repple Depple, and for many of the lower ranking permanent party "EMs". However, in order to isolate the Military Police from the rest of the station permanent party the 1257th had this barrack all to its self. Thus our small contingent was able to "spread out" with no more than two men per room. So in our rooms there was only a single two-foot wide bunk next to each of the six-foot walls. Next to the head of each bunk, a two-foot wide three-drawer dresser was attached to the outer wall. Between the two dressers, there was a steam radiator against the wall and space for a chair. Beside the foot of the bunks, on either side of the door, there was a built-in narrow clothes and storage locker. The floor, as well as most of the barrack floor, was covered with a dark brown asphalt tiles. Above the head of each bunk, there was a four-pane casement window, which was topped by a two-pane transom window. The windows were covered at night with a blackout curtain. We were allowed some freedom on how we decorated our rooms as long as we kept them neat and clean. For example, I had obtained a piece of plywood and installed it between the two dressers to make a desktop. I painted the desktop robin's egg blue. Tools, hardware and paint were readily available in Stone.

MILTON COOK

MP M. E. COOK

> A photograph of me in the summer duty uniform; to send to the relatives and friends back home. It was a war—time custom to have and to share photos of family members who were far from home serving in the Armed Forces. A copy of this particular photo was "returned" to me by the daughter of one of our adjacent neighbors. My quarters were in the room that was directly behind the spot where I was standing. The utility room was on the other side of the brick wall.

From the window of my room, I could see the back of the main Howard Hall facilities building. On the left a wing contained the enlisted men's lounge with the American Red Cross Canteen snack bar. Beyond the top of the facilities building, the roof of the station—theater was clearly visible. The building continued to the right to become the back wall of the mess hall. A row of offices and a wide passageway were located in the middle structure between the mess hall and the theater. On around past the mess hall the street ran up a slight incline, between the two facing rows of barracks, and came to an end at the steam boiler building. The barracks were set back from the street, and except for the paved walkways, the open space between the walkways was covered with furnace cinders. Located on the other side of the street, where it ran around the end of the mess hall complex, were station functions such as the barber-shop, the post office, the dispensary and the PX (post-exchange).

1. In front of MP barrack, common room at rear, right wing where I was quartered is not shown. Cook, Saloka, Andriko (?).

2. Co-inductees Cook, Foiles, Collins. Rear of MP barrack and common room.

3. Rows of barracks and steam boiler building. View is looking to left from vantage point 1.

4. View from right wing of MP barrack Theater at left. H.Q. offices in center. Back of mess hall & Officer mess on the right.

5. Permanent party officers mess and back of mess hall kitchen.

6. Permanent party officer's mess and other buildings.

VIEWS OF HOWARD HALL

Chapter 13

Required Credentials

A POST EXCHANGE

> One counter, that was all there was in the quintessential Repple Depple Post Exchange. The counter might some times be watched over by more than one GI, including WACS.

One of the documents that I always carried in my wallet was a PX ration card. Each week we were allowed to purchase seven packages of cigarettes or twenty-one cigars and two boxes of matches.[15] I would buy my allowed cigarettes and matches for someone else who did smoke. The cigarettes were a nickel per pack, and who ever I got them for would pay me my out of pocket cost. The weekly ration included six ounces

---

[15] These were the so-called penny boxes of matches.

MILTON COOK

of candy, which was frequently made in England Mars bars or Cadbury bars. I would also get the allowed pack of gum or a roll of Life-Savers. The ration also included two razor blades and one bar of soap. I shaved infrequently because I had a very-light beard and so seldom got the razor blades. And I did not know anyone who used a bar of soap a week.

PX RATION CARD

> The PX Ration Card allowed a choice of gum or of Life Savers, but some times both would be available. Although the expression "Any gum chum?" might some times be heard, no youth ever exclaimed "Any Life Savers soldier?". It is unlikely that any GI would have to use a bar of soap a week, but who ever established the ration must have considered that it was reasonable. But the real enigma was the seven pints of beer. Any beer would have been locally brewed draft, which was typically served in a glass or a mug. Now if the PX did in fact sell beer was the GI allowed all seven pints on the spot, as were all of the other items in that weeks ration; and if not just how was the ration to be tallied during the week? When any one could go to pub in town and swill their fill was this ration to just be applied at the station Officer and NCO club bars?

Other important documents that I would normally carry with me were my Soldier's Individual Pay Record with the attached immunization Register, the European Theater of Operations, U.S. Army identification card, and a Class "B" Pass. The Soldier's Individual Pay Record along, with dog-tags, was the main means of identification. The enlisted man's identification card was issued in the ETO, probably, to provide a means for the British authorities to identify the American service man; a quasi passport. This assumption is based on the fact that my weight was also recorded in Stones and pounds. The singular British system of measurement, that is only applicable to members of the human race. The Stone is equal to fourteen pounds, which is somewhat strange and arcane; unless of course you are British. Even though this identification card was applicable to the entire ETO, Stones are also used to represent body weight in some Mediterranean areas. A Class "B" Pass, except for some restrictions, allowed me to leave the station when I did not have any duties to perform.

US ARMY IDENTIFICATION CARD

Apparently the I.D. card was in the purview of that third branch of the Army, the Service of Supply (SOS).

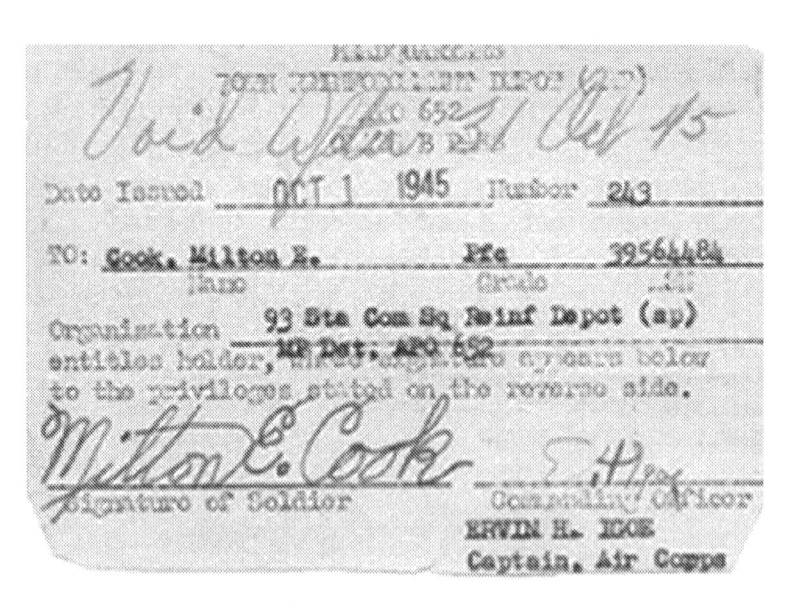

CLASS "B" PASS

This is the last pass issued while the company was still quartered in Howard Hall.

Chapter 14

Settling In For Multiplex Duty

Not knowing when this war would end was the on going basis of constant low level futile fretting. Boredom and the indefinite, was our way of life as we engaged in endless days and nights of MP activities. Many of our activities were regular routines; such as required military formations, cleaning our rooms and the barrack, maintaining our person and our uniforms, and our entertainment. Our actual MP duties were manifold, and the assignments were varied, which at least kept things somewhat interesting. At my level, I was never completely aware of what everyone else in the company was doing and only knew or for that mater cared what was happening relative to my current assignments.

After the regular workday had ended and for a reasonable time after the pubs had closed at twenty hundred hours the casual GIs were allowed to be off the stations on a pass. During this time MP foot patrols were maintained in the four local communities that were within walking distance of the stations. I was on occasion assigned to ride in the vehicular patrol that supported and retrieved the foot patrols. During the regular work-day each of the three station gates were each manned by two MPs. Two of the gates and the guardhouse had men on duty around the clock. A small contingent of the 1257th was assigned to and performed gate guard duty at a close by station that was a part of the local Repple Depple complex. There was just one critical intersection where an MP was posted on point control during the regular work-day and while soldiers out on a pass were going to and returning back from town. The company was to march in two Battle of Britain parades. On a number of occasions, I was set off or sent out on assignments on just my own responsibility. There were other activities such as motorcycle patrol and road-blocks that I was not involved in at any time.

For the most part, patrol duty and gate duty was performed by pairs of MPs. Men who were from the same platoon and squad, and that were also quartered in the same wing and, in some cases, the same room in the barrack. The soldiers whom I was usually paired with, Sgt. Foiles, Cpl. Strehlaw, PFC Amrhein, and PFC Szucs, were all fellow members of the 1257th from the first day that it was assembled. Frequently, when we would be off duty at the same time, we might go out on a pass or otherwise do something together because we were all buddies.

Chapter 15

Rambling on Foot Patrol

When the assignment was foot patrol duty, we would attire ourselves in the appropriate military police uniform for that particular night's deployment. The weather was a major factor in the selection of which items of our uniforms that we would wear. There was some latitude as to the configuration, which could be worn with the imperative that the uniform elements of each pair of MPs be identical. After we were in uniform, and depending on the weather, would assemble in or out side of the guardhouse to have the night's assignment prescribed. It was then we would find out whom in our squad we would be partnering with and in which community we would be on patrol-duty for that night. There were four communities for which the 1257th provided the MP function. These communities ranged in size from Stafford, a fair sized city; Stone, a mainline town; Eccleshall, a vintage village; to the smallest, Millmeece, a country hamlet.

Then we were on our way, scaling up the back of an two and half ton army truck to sit on two facing fold-down slat benches, which were hinged along the sides of the truck bed. We were more or less protected from the wind and rain and fog by a canvas top, which was pulled snug over a series of arched strips inserted in sockets along the sides of the truck bed. The canvas cover was attached to the front and to the sides of the truck bed. At the back of the truck, a foot of hemmed edge canvas was folded around the arched end strip and pulled tight to form an arched opening much like a covered wagon. We had a view of the just passed English countryside through the opening over the closed tailgate. Those seated on the rear end of the benches had the best views. If seated near the front of the truck bed, you could see only a little view of the countryside, way off in the distance. We would pass a sign that provided the information that Izaak Walton's Cottage was down a lane that we had just passed. At that time, fresh water sport fishermen were sometimes referred to, by the press, as "Izaak Walton's" because of Walton's Books on this subject. As the truck wandered through the countryside on its way to our various duty stations, we would loudly sing the marching songs, which we had sung during our basic training. Those who knew the words would sing some English drinking songs that they had picked up while hanging out in the pubs. When the truck stopped at a duty station the assigned MPs would swing over the tailgate and drop to the pavement.

Chapter 16

Rambling on Patrol-Stafford

Stafford was the county seat and was a large enough city to have a number of multiple story buildings. In the downtown area, there were several half-framed buildings with thatched roofs surviving from medieval times. Being the largest community in which we patrolled, there were more pubs for GIs to hang out in, and so there were more pubs to visit and thus keep us better occupied. There was a social club that held dances and provided entertainment for some of the GIs and local girls. We would show up several times a night just to watch the dancers and converse with anyone who would talk to us just to kill a little more time. Even with all of the pubs and the dance, we still had time on our hands. We would amble around exploring that part of the city that was not too far from our assigned area. A visit to the railroad station for a cup of tea would occupy us for a little while, not that there were ever any U.S. soldiers on the platform. After ten o'clock, when all of the pubs had closed their doors and all of the GIs had scattered, the jeep patrol would pick us up, and we would ride back to the guardhouse at Howard Hall.

In addition to the regular MP patrols, Stafford was the location of a significant event in the annals of the 1257th. It was early on a September morn that a major contingent of the company was dispatched to Stafford. The 1257th had been honored by a request to march in the annual Battle of Britain victory parade. This event would be the companies second, longest, last and in some ways the most memorable public parade in England. The particulars of the two Battle of Britain of Britain parades, in which the 1257th marched, are a subject that is covered in detail later on.

Chapter 17

Rambling on Patrol-Stone

Stone had one major street that descended through the town. There was an outlying pub just past where the road from Yarnfield connected to the main highway going into Stone. This pub would always waylay some of the soldiers out on a pass, who would then never make it into Stone proper. Continuing on past the pub, the highway made a rounding left turn into the center of Stone and its main street. To the left, the main street ran up the grade to its end at the parking lot in front of the railroad station. To the right, High street continued down the slope and turned right at the bottom of the grade. A second street made its way up the slope at about thirty degrees from the open area where the highway entered in to Stone. The "Y" intersection of streets formed the town triangle, which was watched over by a bronze WW I Tommie. The right forking road continued up the hill to a residential area. Just a short way up this street, on the left side, was the location of the American Red Cross Canteen. The Canteen provided coffee and donuts and was a good place to hang out and kill time.

At the top of the grade, where the main street ended at the railroad station, a large circus type tent had been erected on an open area between the parking lot and the railroad tracks. This tent was the location of Stone Baggage, which handled all of the gear of the Air Force personal passing through the local Repple Depple stations. Stone Baggage was infamous throughout the Army Air Force in the ETO for pilfered luggage. It was just by Stone Baggage one night while we were on patrol that we heard of the death of President Roosevelt.

THAT'S THE WAY THE BALL BOUNCES

AMERICAN RED CROSS CANTEEN   HIGH STREET

An open-air pedestrian bridge was used to cross over the railroad tracks to the opposite platform. A pathway from this platform made its way to a lock on the local artery of the country-wide canal system. A canal system that had fostered and spread the industrial revolution throughout Britain; which had long since been supplanted by the railroad network. Never the less, there were still canal boats with families living on them plying the canals. A way of life that more than likely would have already vanished, had it not been for the slump and the war. I assumed that these boats might have been aiding the war effort by transporting some type of bulk cargo. Today, all of the remaining canal boats have been converted into houseboats, and much of the canal network is still maintained for recreation.

Returning past the parking lot in front of the railroad station, on the left, at the very beginning of the main street, was also the first pub. A pub that is still "serving 'em up" to this very day at this good location. About half way down the grade, to the town triangle and on the right, was the Catholic Church that held dances for the GIs and the local girls.

On past the triangle, the main street sloped on down through the heart of town. The narrow street was lined on both sides with shops, hotels, restaurants, fish and chip shops, and pubs. Part of the way down the street on the right, there was an open-air marketplace. A little farther down the street, still on the right, was the location of the

Stone brewery. The bouquet of brewing bitter British beer borne on the breeze blended into the background of Stone. Alas, the Stone Brewery is no more. Today, its old home is occupied by a supermarket and an arcade of facing shops on a passageway to the parking lot and the highway, which now bypasses the towns main street.

At the bottom of the slope, the main street made a sharp turn to the right and shortly departed from Stone via a bridge over the lower level of the canal. The last pub as you were leaving town was on the right. This pub had a view overlooking the last lock in town that completed the canals stair stepping down the grade, as it ran past Stone.

There was much excess time, as usual, to dispose of when patrolling up and down through Stone. There were only a few pubs, and we could only linger a few minutes in each. They were usually smoky and so filled with people that it was difficult to even get into the places, so it was just in and out of the pubs. A stop at the American Red Cross Canteen for coffee and a donut would only take so much time, and it was something that should not be done too often. As usual, we would explore up the side streets as far as we considered reasonable. We would fritter away some time looking again at the displays in the shop-front windows. One shop window displayed a very old military decoration from Oliver Cromwell's tenure in the early seventeenth century. The figure on the silver medallion hanging from a ribbon was almost worn away from many years of polishing.

After the pubs had closed at ten o'clock, we would encourage the GIs to be on their way back to Yarnfield. When we had herded all of the soldiers out of Stone and they were all on the road on the way back to the station, we would get into the patrol jeep and ride back to Howard Hall.

CHAPTER 18

Rambling on Patrol-Eccleshall

Eccleshall dated from at least Roman times, and was most likely located at the road junction, even as it is today. Most of what we had to be concerned with as MPs was located on two "T" intersecting main streets. A cornered Norman style church was situated on the right side of the major street at the far end from where we entered the village. The center of the village where the pubs and shops were located was relatively flat with only a slight grade going up the intersecting main street. Housing complexes (Halls), similar to those that comprised the Repple Depple, faced each other across the road a short way before it entered Eccleshall. Both of the housing units were the residence of a multitude of young women who were employed in the nearby munitions factories.

This being a very old village, some of the buildings also had to be very old. The buildings were predominantly red brick while the church, naturally, was constructed of gray stone. One of the old pubs, near the end of the intersecting main street, had an uncommonly large room. An enormous hand formed tree trunk of a beam bisected the ceiling of the room to support the floor above. It was in this room one night that a group of inebriated paratroopers were singing one of their marching songs. This unusual song went something like this, "Oh, it's stand up and hook up and hope your chute won't fail." This song was the first clue, and their shoulder patches confirmed, that these soldiers were members of an U.S. Army Airborne Division. It was strange to see any GIs who were not members of the U.S. Army Air Force on pass from the Repple Depple. I thought maybe these paratroopers were away from their usual haunts and must have been involved locally in some training exercise.

Dissipating time was even more of a vexation in Eccleshall because the village was small, and there were only a limited number of pubs that should be visited—not too often and not too long. One of the things we would do without fail was to stop in a tobacco shop located on the right side of the street soon after entering the village. In this shop, we would spend some time visiting with the proprietor, our friend Ellen Hurlstone. As I remember, she was a retired schoolteacher. Through the open rear door, you could see into her home, which was part of the same building as the shop. The panes in the front windows and door of the shop were old style bull's-eye glass panes. Frequently, when we had stopped in her shop for a visit, some GI would open the door and say, "Is this a pub?" To which we would all reply, "No, the pubs are on down the street." At that time of the day, her shop was the first open business when you entered the village.

For Christmas 1944, Ellen Hurlstone gave gifts of a small volume of one of Shakespeare's plays. The volume of *Twelfth Night* that she gave to me is two inches tall, by an inch and three-eighths wide, by three eighths of an inch thick. I could tell that it had been a volume from a set that she had used for gifts, as the color of ink used for my name was blue, and the color of the ink for the rest of the inscription was black.

I do not recall, but Ellen Hurlstone had to have asked for my mother's address. I have a packet of post cards, which she had sent to my mother with her comments on the back of two of them. My sister, Joy, in turn sent Ellen Hurlstone a packet of post cards with scenes of California. Then in September 1945, Ellen Hurlstone sent my sister two small pieces of old Staffordshire china with a letter in which she said that she had not seen me for some time, but when she did, she would let her know about how I was doing.[16] But I was never to see her again as we had stopped patrolling in Eccleshall and would depart for Germany in early October.

After my sister receive the China, my mother asked when she next wrote to me, was there any small items that could be sent to Ellen Hurlstone in appreciation. My reply was to send her some linen

---

[16] See endnote 4 for comments on the post cards and for Ellen Hurlstone's letter.

handkerchiefs. They were impossible to obtain in wartime Britain because clothing was rationed. However, the new handkerchiefs should be washed before they were put in the mail. We had been informed that any new clothing sent from the US to the UK should be washed (there by making it "used"), before mailing it, to avoid paying a stiff customs duty.

Eccleshall being a village, any street along which we might walk would soon deliver us to the countryside. To fill the surfeit of time between the spaced checking on the pubs, we would sometimes stop by the Police Station and spend time conversing with the "Bobbies." But most of the time, we would wander around from place to place and engage in various mental diversions. We would entertain ourselves by trying to remember and sing the popular songs that we had known back in the States. For the most part, as members of a platoon, we had already told each other all that we needed to of what had transpired in our short lives up to this point. There was little discussion of the war as most of our information came from the Stars and Stripes, and there was not much to be gained rehashing what we all knew.

Being placed in a position, as a young and unsophisticated individual, where your function is to be an observer of your fellow man, you become inclined towards a singular view of life. In this position, you are regularly exposed to all of the assorted follies and foibles accomplished by your fellow man. The really dumb and foolish happenings that we observed would sometimes amuse us and sometime appall us. With all of these instances of strange and unusual "conduct", you had to develop a philosophy of thinking first before engaging in what might be foolish behavior. We would hone this philosophy during our discussions as we rambled along the various streets through the various towns on or various patrol activities. And thus, we were more able to avoid post-traumatic stupidity disorder.

During the summer, the clocks were set ahead two hours for double British Summertime. At the latitude of the English midlands, with the two-hour advance on the clocks, it was still light quite late at night, which would hamper some activities. In the winter, the clocks were only set forward one hour for British Wartime Wintertime, which could be construed as British Summer Time during the wintertime. Some times during the winter, when the temperature was near freezing, the relative humidity would be close to one hundred percent; a marrow congealing cold and dampness. After a while, as a result of going into the hot pubs and then going back into the cold night air, our clothing would become damp and lose its insulation properties. There was a humidity chill factor, and I have heard many of an ex-GI say that the winters in England were the coldest that they had ever been in their lives! No matter how much clothing you had put on, you would become miserably cold, and then you could not get warm while being inside. We would wear our field jackets under our overcoats and a pair of cotton fatigue trousers under our wool trousers, which helped a little to keep us warm. We had been issued gray RAF scarves, because we had to spend so much time out in the cold. And these RAF scarves were our only official winter clothing. Some of us would eventually have hand knit sweaters and sweater vests that had been sent to us by our mothers. But at least when on patrol duty, we were moving and did not get cold feet.

    Herding the sheep out of Eccleshall took more time than usual. Getting all of the soldiers headed back to the Repple Depple took required more attention, as they tended to tarry with the women who lived in the munitions workers housing complexes on both sides of the road. We, of course, would spend the time talking to some of the young ladies until the last straggler was past temptation. Then, it was into the patrol jeep and back to Howard Hall. The motor pool had enclosed some jeeps with Plexiglas and plywood sides and doors, which in the winter made this late-night trip, a little less of an ordeal. On nights when the patrol jeep returned to the guardhouse with all of the patrols packed in, we would spill out of it like a bunch of circus clowns.

Chapter 19

Rambling on Patrol-Millmeece

Millmeece was the home of the munitions factories, long gray buildings with saw-tooth roofs rising slowly on the north side and dropping swiftly on the south side. This aggregation of buildings, roads, power lines and railroad tracks was guarded by a high cyclone fence, which was topped with spiraling barbwire. Eleanor Roosevelt had visited these munitions factories not long before the company had arrived in England, so we were told. The only isolated local station of the Repple Depple, Nelson Hall was also situated close by Millmeece. The proximity of this Air Force Station to the small country village resulted in a large influx of soldiers into the few pubs and restaurants, which in turn required the presence of MPs on patrol.

The limited size of the village and the scant number of pubs meant that there were even fewer things of interest to keep us occupied when patrolling in Millmeece. During the summertime, when it was still light late in the evening, we could walk up a road to the top of a little hill. At the top of the grade on one side of the road, the land sloped off with a vista of rolling green countryside interspersed with red brick farm buildings. On the opposite side of the road, on slowly rising ground, there was a gated country estate. The manor house was backed up against the estate's expansive woodlands. In the foreground, at a slightly lower level, were the stone remnants of some ancient edifice sprawled across the ground. After ten o'clock, when the pubs had closed and the last GI was on his way back to Nelson Hall, the patrol jeep would finally pick us up, and we would pack into it and head back to Howard Hall.

Chapter 20

Rambling on Patrol-Back At the Guardhouse

In the far corner behind the cell, diagonal from the entrance to the guardhouse, there was a small kitchen. This little "mess hall" was as wide as and in line with the guard-room latrine. There was a small counter top, on the lower half of the split door, on which the food and coffee were served. And uniquely the coffee was served in mugs with handles, no canteen cups at this mess. You had to go outside and walk around the guardhouse to get to this mini kitchen with small appliances and a large coffee maker. Someone on duty at the guardhouse would also have the job of having a full pot of coffee and fixing the snack for the returning patrols. The snacks were sandwiches made from whatever the cook in the mess hall had to send down to the guardhouse for our snack. I always thought that the pork chop-with-the-bone-still-on sandwiches were the most interesting. There was a steel picnic table with attached benches available, but most of the time after we returned from patrol duty, it was too cold to sit down.

Captain Bishop, the Company Commander, the Provost Marshall was a dapper officer. He obtained leather Sam Brown pistol belts for the whole company, maybe because he wanted one for himself. At any rate, he socialized with some of the local gentry, and apparently, joined in the hunt on their estate. However, I think that he was out of the loop, and he was not aware of our snack when coming off patrol duty. One day he wanted to know what had happened to his rack of venison, which he had left in the kitchen. Of course, we had no idea what he was talking about. No one knew anything about a rack of venison. We did not know just what the cook has sent to us, but it was unusually tasty!

CHAPTER 21

The Way of The WACs

Isolated between the guardhouse and the theater, there was a long rectangular barrack building. This building may have been intended as housing for some administrative personnel when Howard Hall was planned as housing for the munitions factory workers. This building was used as the permanent party officers' quarters. But, with the arrival of a company of Women's Army Corps soldiers, the building was ideal for their barrack.

The WACs were the only unit at the Repple Depple that routinely marched as a company. Typically, they were the only unit that marched to the Mess Hall. And, typical of U.S. soldiers, they sang as the marched to such songs as "Be kind to your web-footed friends, for a duck may be somebody's mother." Most of the women were employed in the major function of the Repple Depple—that of processing the paperwork of the Air Force personnel arriving from and returning to the continental United States. A few of the WACs performed other duties such as working in the PX.

There were several marriages of WACs to other soldiers on the station during their time together in England. Events, which at that exact point in time, instigated a brand new meaning for a hitch in the Army.

CHAPTER 22

Occurrences In the Barrack-Best Foot Forward

One of the things that attracted my attention and somewhat fascinated me when we arrived in Great Britain was the glistening brass. Door handles, letter drops, fittings and decorations on horse bridles and tack, military buttons, badges and buckles and any thing else made from brass were each and every one shining brightly. Keeping the bright work glowing is an old nautical custom and for a long time a vast body of British seafarers must have brought this practice with them when they came home from the sea.

The United States Army winter service jacket had four brass buttons down the front closure, smaller brass buttons on each of the four pocket flaps and on the two epaulettes.[17] One-inch diameter brass ornaments were affixed to the upper collar flaps. The raised letters U.S. were displayed on the right side disk. A pair of crossed muzzle loading pistols—the insignia of the Corps of Military Police—was attached to the disk on the left collar. A larger diameter brass ornament with a raised Army Coat of Arms[18] was attached in the crown of the peaked service cap. When issued all of these brass buttons and devices were coated with lacquer that gave them a dull yellow appearance without a hint of a shine.

---

[17] The American Army used handkerchiefs, so there was no need for buttons on the coat sleeves.

[18] The Army coat of arms depicted an American eagle with Stars and Stripes set on a shield on its chest, the right claw holding a sprig of laurel and the left claw holding a bunch of arrows. The coat of arms was also on all of the brass buttons and on the olive colored plastic buttons on the overcoat.

THAT'S THE WAY THE BALL BOUNCES

The British manufacturers and merchants of military fittings and finery ever eager to make a quid tapped into the expanding U.S. Army market for supplements and appendages to their uniforms. The service cap which was an authorized part of the uniform, but which had not been issued, was acquired from this British source by almost everyone in the company. Unit shoulder patches, badges of rank—chevrons—and overseas bars were readily available and the British tailors were ready to sew them onto uniforms. Interestingly, the design employed on English-made Eighth Air Force shoulder patch had shorter and different shaped wings than those embroidered on the American-made patch, and the difference was very obvious. I, myself, bought an American type brass whistle and chain to use as a uniform decoration when I was on duty. Brass uniform buttons in the British style, that is the Army Coat of Arms without the raised outer ring and background ribbing on the standard army buttons, were available. Some of the more flashy members of the company who wanted to further personalize their uniforms made the most of it, and opted for these rimless buttons. Brass was of course a critical war material used for shell casings, which were later made of steel to free up the brass for military uniforms.

EIGHTH AIR FORCE SHOLDER PATCHES

American made patch on the left. British made patch on the right. This patch is still in use on displays of The American Air Museum in Britain.

Observing all of the bright British brass and being disposed to putting on a show from our training, it was only a natural reaction that the 1257th would start polishing all of our own brass. We were already applying British whiting to our leggings and our web belts, so shining our brass was only the next small step. Brass polish was an English staple, and so also the uniform guards and short bristle brushes used by the British military for shining the buttons on uniforms were essentials. The uniform guard kept the uniform clean when polish was applied to the buttons that were then brushed and wiped clean. The guard was a one thirty-second inch thick sheet of brass about one and one half inches wide by about six and one half inches long. There was a five-inch long by one-quarter inch wide rounded end slot from one end of the guard. The three free ends of the guard were circular shaped. In use, the guard was slipped under a button to the end of the slot. The fingers were slid under the guard with the thumb resting against the side of the button. Polish was dabbed on the button, which was then brushed briskly. The button was wiped clean, and the guard was pulled from under the button. After a few sessions of polishing our uniform buttons, all of the lacquer was worn off, and then they gleamed like white gold. Once started down this path, there was no turning back, so we shined, and were expected to shine in addition to our buttons, all of our brass uniform ornaments, buckles, fittings, whistles and even our dog tags.

All of our large articles of clothing and other personal gear were identified by marking them with the first initial of our last name followed by the last four digits of our army serial number. Early in basic training we had been advised to and did buy a large rubber stamp of our identifier and a black indelible ink stamp pad. The rubber stamp provided a quick and neat way of marking clothing and gear with our identifier. The identifier was 4484 (the last four digits of my Army serial number) in my case, and I can never forget it! All of our cotton clothing went to the base laundry for washing. The small things such as socks and handkerchiefs that could not be marked were enclosed in large cord mesh bags, which were then tagged with our identifier. Our wool uniforms were sent someplace for dry cleaning. These wool uniforms would always come back from dry cleaning with an aromatic odder suggestive of high-test aviation gasoline. However, the dry cleaner's pressing of the uniforms was not up to our high standards. To remedy this pressing problem and to maintain our high standards, we would usually every day before going on duty re-press our uniforms. In the utility room, a large table covered with an Army blanket served as our ironing board. Someone had obtained a tailor shop sized electric iron which, using a moist handkerchief, was used to steam press our uniform smooth and make all of the crease edges sharp enough to cut before we went on duty.

The brown high top U.S. Army shoes were always replaced when a hole had been worn in a sole, which happened with some regularity. Sometimes, like the replacement shoes would be strange. One such unusual replacement issue was the inside-out shoe. On these shoes, the rough side of the leather upper—the inside of the cowhide—was on the outside of the shoe and the smooth side of the leather—the outside of the cowhide—was on the inside of the shoe. This was a significant problem for any recipient of a pair of these inside-out shoes, as it was a cultural imperative and quasi dress code requirement that we must keep all of our leatherwork glossy. The problem was more or less resolved by applying many, many layers of shoe polish until a relatively smooth and shining surface resulted. Even stranger, one of my comrades was issued a pair of shoes with hobnail festooned soles. Now, soldiers in the British Army had hobnails on their boots, but the U.S. Army normally eschewed hobnails on their shoes. A hobnail is a small iron protuberance attached to the bottom of the shoe by two tabs that were punched through the bottom layer of the sole and then bent over to hold the nail in place. It is reasonable that unless ordered to, no one would want hobnails on his shoes as they make a terrible racket and are slippery when walking on hard surfaces. I had the unlucky soldier who was issued this pair of shoes, put them on and then rest the toe on a chair with the sole vertical. Then, using my hammer and screwdriver, I popped every one of those little hobnails off of the soles of his shoes.

Sometime in 1944, we saw the advent of the combat boot. The standard high top Army shoe was promoted to combat boot by sewing a leather cuff onto the top of the shoe. After the shoe was laced up and tied, the added cuff was closed using two straps and brass buckles on the outer side of the cuff. The best thing about adding the cuff to the Army shoe was that it would lead to the expungement of the dismounted leggings. It would be a drawn out affair to switch from our white leggings to combat boots. The change over would create a minor problem as a pair of MPs on duty had to wear identical uniforms. We did speed up the changeover by removing the cuffs from worn out and discarded combat boots and sewing them onto the tops of our Army shoes and thus converting them into combat boots.

In a discard pile someplace, probably Camp Miles Standish, I had found a pair of brown dress oxfords in my size. This was a real bit of luck, so whenever I went on a pass I did not have to wear my high top GI shoes, but I could wear my good shoes. And then, I received, in the mail, a present of a pair of brown half-Wellington boots from my buddy, Vernon. These boots had been made in Mexico, and Vernon picked them up in Texas where he was training at the time. Now, I had even better and warmer foot-gear to wear whenever I went on a pass.

Keeping our shoes and other leather articles shining was as much of our culture as keeping our brass work shining. I bought a British shoe-shine brush and used it on all of my leather gear. The American shoeshine brush has longer and softer bristles than the English shoe-shine brush. The English brush performs a much better job than the American brush when it comes to eradicating scuff-marks from shined surface of shoes. Using the English brush, it is very easy to work the shoe polish to fill in the scuff-mark without any added polish. I still use the same English shoe brush that I obtained in 1943, if I ever feel the need to shine my shoes. This brush still has an eyelet, used to hang it up in my locker, on one end that was made by driving a finish nail into one end and then bending it over to form a loop. We always used the very best English shoe polish on our leather equipment. And yes, we did know the difference between "the other stuff and Shineola."[19]

---

[19] Credit this to Betty White, who spoke thus on the TV show *The Golden Girls*.

CHAPTER 23

Dining in the Mess Hall-Holiday Feasts

The United States of America military establishment mounts a major logistic thrust to serve a traditional dinner to all of its armed forces wherever their location may be on Thanksgiving Day and on Christmas Day. On the standard menu for this banquet must always be roast turkey and fixings. The usual fixings, as you might expect, included some type of dressing, mashed potatoes, giblet gravy, candied yams (sweet potatoes from the can), some kind of canned vegetable, cranberry sauce and pumpkin pie. Before returning to the United States of America, I was to partake of this repast deposited in my mess kit six times. We were served the turkey dinner in England on both holidays in 1943 and 1944. Then, in 1945, the Thanksgiving turkey feast was in Germany and the Christmas turkey feast was in France where we were staged for the trip home.

A CHOW LINE

Soldiers advancing on the chow line with mess kits and canteen cups deployed and ready for this meals action.

Chapter 24

Dining in the Mess Hall-Some Good Food and Some Bad

    Much of the food that we were provided was high in carbohydrates or as it was called in those times, starches. Eating a high carbohydrate diet, we all got a little fatter. The pictures that I have from this time of myself and some of my comrades show all of us with rounded faces. We were active which controlled the weight gain, and also a little fat may have helped to keep us warmer during the winter. The English wartime bread, which was purchased from a local bakery, was a multigrain loaf with a gray cast and a coarse texture. However, it was tasty and today would probably command a premium price as a specialty loaf. Powdered eggs I have touched on before, but there were also dehydrated potatoes, carrots and onions used by the cooks to produce the food which they then served to us in the Mess Hall. With the morning meal, canned grapefruit or other canned fruit would frequently be served, and of course, some form of fish was on the menu on Friday. One canned meat, which I thought, was really appetizing was the baked ham with raisin sauce. Sometimes, for breakfast we might be served that traditional treat, known to so many service men in so many wars, SOS (stuff on a shingle). SOS being browned particles of ground beef in gravy dumped over toast. SOS was of course a military operation to use up stale bread; that also required only a minimal effort by the cooks and was cheap.

    There were some items on the menu which I could not stomach, and so I avoided them whenever possible. One such food was canned string beans because to me, they had a very strong metallic taste. In fact, I will not eat canned string beans even today because of the metallic taste. On some days, the cook would make up a batch of beef stew. Sometimes the batch of beef stew would be distinctive in that the chunks of beef were, in fact, chunks of fat with just an occasional shred of lean. It did occur to me at the time that lean roast beef or steak was most likely on the menu in the Permanent Party Commissioned Officers Mess on that very same day.

On some occasions, later in and after the war was over, our chow would be prepared from the ten-in-one rations. These rations were supplied in a large water-proof cardboard box that contained all of the food needed to fix a meal for ten men. The ten-in-one rations were made up to supply field kitchens, where a box would be opened for every ten men who were being fed that meal. Now, it may have been that there was a surplus amount of ten-in-one rations that had to be used up, or that our regular supply was running short or maybe some of both.

Chapter 25

Singular Service-Left Out in the Cold

It was a typically cold and dreary mid-winter day late in 1943. Cook was sitting on the bunk shining shoes when Mc Fall, who had day room duty, opened the door and popped in to the room. "Hay, Cookie, the Desk Sergeant called and said to tell you that he wants to see you right away at the guardhouse."

"What does he want to see me about?"

"I don't know. He just said to have you come down to the Guard House."

Private Cook was never in trouble, so he knew it was not something that would really concern him personally.

Cook put on his field jacket and hat and then walked on down to the guardhouse. Entering the guardroom he halted in front of the Desk Sergeant. "What do you want to see me about, Sergeant?"

"Well, Cook, instead of patrol duty tonight I have a special assignment for you this afternoon. There is going to be a troop movement, and there is some concern that the convoy might get lost in the mesh of country roads. I'm going to going to post you on point control at one of the confusing intersections, so in case the convoy does get lost, you can direct them to the correct road. Eat early chow, get dressed for duty and be back here at twelve hundred hours."

"Okay" Cook replied, and then left the guardhouse. Returning to the barrack, he picked up his mess kit from its hook in the locker. Then back in the mess hall, which he had just passed on his way to the barrack from the guardhouse, ate early lunch and returned to his room.

It was cold this time of the year—in the thirties—and Cook, along with the rest of the company, had not been issued any winter uniform clothing. So he took off his wool OD trousers and put on a pair of green cotton fatigue trousers. Next, he pulled a second pair of cotton socks over the pair that he was wearing. The tops of the second pair of socks were pulled over the taper folded ends of the fatigue trousers. Then, pulled his wool trousers back on over the fatigue trousers; put on the high top army shoes; slid the strap at the bottom of leggings over the toes of his shoes and laced them up while tucking the folded ends of his wool trousers into their tops. Then he put on a field jacket, the gray wool RAF scarf, and a wool overcoat. He had to re-button his MP bassard to provide a snug fit on the left sleeve of the overcoat. The length of the web belt had to be adjusted and to fit the side arm around the combined field jacket and overcoat. As they had never been issued he did not have a wool cap to put on before donning the helmet liner.

Cook walked back on down to the guardhouse. He entered the guardroom saying, "Sergeant, here I am ready to go."

"Okay, Cook, get in the jeep, and Bouton will take you to your post." And then they were off to only Bouton knew where.

The jeep traveled along the winding country roads for some time. Bouton stopped the jeep at about the midpoint of a quarter of a mile of black asphalt road, which dead ended on a crossroad at each end. The road was located in the bottom of a long shallow depression. The ground sloped up on both of the sides of the three roads, to a height above Cook's head. Tall hedgerows, maybe six-feet, were growing on the crests of all of the slopes. The fifty or so feet of sloping ground, from the asphalt pavement to the hedgerows, had a few patches of grass scattered over the gravelly soil, not good farmland at all. There was not a farm-house or a tree to be seen from Cook's post, just bare asphalt, the few clumps of grass on the other wise bare-ground and the hedge rows.

"You can get out here, Cook. This is your post. If the convoy shows up, tell them to take that (pointing) road to the right."

"When will someone be back for me?"

"I or someone else be back for you as soon as we receive word that the troop movement has been completed." And with that, Bouton let the clutch out and raced off in the jeep.

The sky was gray and featureless; the sun had not been seen in weeks. It was cold—the air was cold, the ground was cold, even the black asphalt pavement was cold. There was no place to sit down. There was not even a place or any thing else there, on which Cook could at least lean.

Cook wondered, *"How long am I going to have to stay here?"* His feet started to get cold. The leather in shoes that are worn almost every day became damp, and as a result the soles provide very little thermal insulation. Cook walked around stomping his feet, but the stomping did not provide any warmth and just hurt his feet. His feet just became colder and colder. *"Damn, my feet hurt! God, my feet hurt! How long have I been in this place?"* Cook did not have a watch, and with the overcast sky, it was impossible to tell how much time had passed, but it seemed like forever.

"Oh man, my feet hurt! I wonder when the driver will be back for me? I wonder if he will ever come back? I wonder if the convoy will show up? That would at least be interesting if the convoy did show up. Boy, I could really use that pair of arctic overshoes right now!" The fact that others could at that time, and would in the future be in a worse state never entered Cook's mind. It is almost always the case that extreme discomfort tends to inhibit philosophic contemplation.

After an interminable duration, without the sight of any other form of life except plant, Rhudus finely drove up and parked the jeep. "Okay, you can get in now Cookie."

"You bet, if I can." Cook managed to get himself into the jeep, and off they went along the winding country roads. "My feet are so cold and hurt so damned bad that I can hardly stand it. I can't wait to get these cold shoes off on my feet."

"Well, we'll soon be back at Howard Hall, and you will be off duty for the night." And soon enough they did arrive back at the guardhouse.

Cook hobbled back to the barrack and into his room. After removing the helmet liner, MP brassard, web belt, overcoat, scarf, field jacket, leggings and shoes, he flopped down on his bunk and propped up his feet. After a while, his feet had warmed up and almost quit hurting, and the feeling in them had returned to almost normal. And now it was suppertime, so he put the shoes back on his still aching feet. Then he, donned his field jacket, grabbed the mess kit and headed for the mess hall.

But as long as he lived, he would never forget this afternoon's sojourn in the English countryside.

CHAPTER 26

Tales From the Guardhouse-Love and War

Sometime in the late sixties, during the Vietnam War, one night I was watching the television comedy show *Rowan and Martin's Laugh In*. In one of the skits I heard Pat Paulsen say, and these are probably not his exact words, "The Hippies today say 'Make love, not war,' but in World War II, we did both." It's very likely that he did not mean both at the same time, though, I was familiar with tales, which are not entirely apocryphal I think, of newly liberated women embracing the advancing GIs with open arms. But in the usual state of affairs, love and war are mutually exclusive propositions.

In the historical account of the British troops in colonial America, it has been noted that they were delighted with the promiscuity of some of the colonial women—and why would anyone think otherwise. Now the situation was reversed with American men and British women instead. The "mills of the gods" do grind slowly, and that is fine. Even so, it was only some of the British women with this proclivity. Increased loose conduct is almost always the case when moral standards decline with the increased social confusion and separation that occurs during wartime. Factors such as many of the British women and all of the American GIs were away from home and family constraints. Then, there were the cultural differences. In any case, the most were nice and proper English girls, which is not to say that some of these did not mess around. There were many romances, both major and minor. There were many British war brides, who would depart from their snug homes and make the trip to the United States sometime after the war was over, and most of the American soldiers had left for home.[20] One

---

[20] See end note 5 for sources on English war brides.

of the members of the 1257th—a very dapper soldier (Sforzo), the personification of an Italian Count, I always thought—was to marry a local girl. I often wondered how that particular combination worked out in the long run. I dated a nice girl, who I would meet in Hanley when out on a pass, but more on that later.

Still in war, as in real life, love could be very strange; take for instance this account of love between a Yank, well two, and a Brit and their some—what unusual relationship. One of the permanent party sergeants managed to get his English girlfriend "knocked up!" This was somewhat of a problem for him as he already had a wife and family back in the good old U.S.A. How would he ever be able to explain to his wife back home, that he was providing child support for some foreign bastard. However the Sergeant, rising to the occasion, devised a cunning resolution to his predicament. It so happened that one of the Sergeant's subordinates was a very homely and, I think, not too bright Private. Whatever the involvement was, the Sergeant prevailed upon the extremely homely soldier to take the Sergeant's pregnant true love to be his wife. And so it was that the Sergeant and the exceedingly homely GI would depart from the station on a pass together to spend sometime with his lover and his wife, respectively. I always wondered if the inordinately homely Buck Private got to sleep with his wife. Just what were the arrangements between these very strange bedfellows? It is an enduring enigma of whatever befell this very curious *ménage à trois*.

At some point, an occasional member of the U.S. Army would lose his edge and resolve that for the moment, pursuit of the war was not as important, for him at least, as time spent with some English darling. This being the case, he would overstay his pass and not return to his unit. Thus, he would be classified absent without leave or AWOL. It is true that a few soldiers who went AWOL were in fact deserters or, in the Army vernacular, had "gone over the hill." Desertion while a soldier was on pass or on leave was determined by the fact that he had made a resolution to never return to the Army at any time or at any place. Desertion from the field of battle (before the enemy) was a different and more serious offense, the celebrated and only case of which I will address later.

In the first case, after a soldier had overstayed his pass or leave for an established period of time, he would be classified as a deserter. Now, some few GIs who were classified as being deserters had met with some misadventure, as the Brits say, and was deceased in some forgotten and hidden place. There were some few, so we believed, who had deserted and had in fact "gone native." They became Englishmen and settled in with some British woman. But the vast majority of the soldiers who were AWOL did intend to return soon and in fact would return when they ran out of money.

Periodically, we would be given a thin four by five inch pocket sized booklet which listed the name and rank of all of the thousand, more or less, of just the Air Force soldiers who were AWOL when the booklet when to press. All of the non-commissioned officer ranks were present, but there were not as many Privates as might be expected as they lacked the funds to be long gone. The highest ranked commissioned officer listed in the book of those who were AWOL was Captain. It is very likely that the AWOL commissioned officers were Bomber Flight Crewmembers. The 70th Replacement Depot, in addition to its activity of processing all of the Army Air Force personnel arriving and departing to the United States, also operated a number of English Country Estates as combat rest units for flight personnel.[21] Rest and Relaxation had yet to be invented. Now, some Airmen may have become impatient and decided to provide their own recreation, consisting of drinking and time spent with a female companion. Going AWOL to be with some "nice" English darling was their only crime—a crime of passion.

We never did check passes or IDs searching for any AWOL who might be listed in the book of those who were AWOL. The GIs with whom we had contact were, with the possible exception of Stafford, were just passing through the Repple Depple and would be gone in a few days. Any soldier who was AWOL would try to stay away from any area where he might encounter MPs, in any case. There is much more area in which to get lost in any larger metropolitan area, but always with the risk that someone will report a lone American soldier just hanging around all of the time.

My first involvement in nabbing an AWOL GI was when I was assigned accompany one of the company's Lieutenants and Sergeants to investigate a report of a possible AWOL. We mounted up in a

---

[21] See endnote 3 for detailed information on the combat rest stations

jeep and drove to one of the towns located in The Potteries[22] and to a home where it had been reported that an American soldier was staying. The Lieutenant said, "Cook, you go on around the house to the back door, just in case our AWOL GI decides to head out that way. If he does, place him under arrest." The Lieutenant and the Sergeant went to the front door of the house. They were invited in and soon came out of the house with the AWOL GI. The Lieutenant ordered me to return from the back of the house and said, "This soldier is in your custody until we return to the Guardhouse." With that, he and the Sergeant went back into the house. I supposed then went back in the home to talk to the GI's girlfriend and her parents. The AWOL soldier asked me if he could tell his girlfriend goodbye, and I, being young and softhearted, said that I would ask to see if he might tell her goodbye. When they came out of the house, I asked if he might tell his girlfriend goodbye, and both the Lieutenant and the Sergeant said, "NO!" I told my prisoner to climb into the left side of to the back seat, and then I got in and sat on his right side as my side arm was on my right side. The Sergeant and Lieutenant got into the jeep, and we made our way back to Howard Hall where the AWOL GI was booked into the Guardhouse cell. I've have often wondered how it all came out. Did those two lovers ever see each other again? If so, how did this particular love affair finely turn out?

---

[22] The Potteries is a metropolitan area comprised of municipalities such as Stoke-on-Trent, Newcastle-Under-Lyne, Stoke-Upon-Trent, Hanley, Fenton, and et.al. It is also the home of Wedgwood, Royal Dalton, Spode, et al. Hence the appellation—"The Potteries".

THE POTTERIES

A long view of The Potteries, or in the vernacular The Podres, as they existed at the time. Centered in the English Midlands, the heart—land of the Industrial Revolution. Where in 1782 Josiah Wedgwood first applied steam power to the manufacturing process when he installed a Boulton & Watt steam engine in his factory.

Chapter 27

Singular Service-Orders to East Anglia

The Desk Sergeant opened the pass window, "Cook, come in the Guardhouse for a minute. I need to talk to you about some thing that has just come up."

"Okay, I'll be right in. Shaw, if you need help give me a call."

"Cook, a special order has been issued for you to go to Station 117 and escort back here a GI, who is being held custody in their Guardhouse."

"Oh, when will I be going? Where the heck is station 117? And how am I going to get there and back?"

"Station 117 is located at Kimbolton, which I think is one of the bomber airbases that are located in East Anglia. And you will be going there and returning on the regularly scheduled trains of the British Railroad system. Your train will depart from Stone Station at zero seven hundred one eight hours, so you be here at the guardhouse by zero six thirty hours, ready to go. So you better hit the Mess Hall in the morning as soon as it is open. Your orders and travel documents will be ready when you get here in the morning."

"Alright, Sergeant I'll be here early tomorrow morning, ready to go." Then, he went on back outside to his post where he remained with Shaw until relieved by Carter and Hiaval who had the next duty shift at the Howard Hall gate.

When Cook arrived back at the barrack, he stopped in the Day Room and told Dunn, who was on duty, that he needed a wake-up call at five hundred hours. Then he went on down the hall to his room and relaxed for a while reading an Armed Services Edition Book.[23]

---

[23] See endnote 6 for discussion on the Armed Services Editions Books.

When it was chow time, he grabbed his mess kit and headed to the Mess Hall with his comrades, most of who would be on patrol duty a little later that night. After returning from the Mess Hall, he picked up his book and read some until it was time to hit the sack.

"Hey, Cookie! Wake up! It's five hundred hours."

"Okay, okay Jackson! I'm up!" And Cook sat up, and as usual, when getting up early from a sound sleep, he felt a little dull. He grabbed some clean underwear and headed down the hall to the latrine, where he relieved himself and took a shower. He did not need to shave this morning. Back in the room, he dressed, being careful not to wake his roommate, put on a field jacket, grabbed the mess kit, and headed for the Mess Hall. After a breakfast of Spam, "scrambled" eggs, toast, canned grapefruit and coffee, he returned to his room to pack for his trip. A set of clean underwear, a pair of socks, a towel, an MP brassard and his toilet kit went into his musette bag, along with a couple of books. Cook took off his field jacket and put on his service jacket, and then hooked his web belt and sidearm around his waist. He was just a little leery about carrying the sidearm in his musette bag, just in case he somehow got separated from the bag. The overcoat would cover and hide the sidearm on the trip to wherever he was going, and then he would wear it on the outside on the return trip with his prisoner. He put on his service cap as a white helmet liner was not suitable for this type of assignment, and he headed down to the Guardhouse.

It was about fifteen minutes after six hundred hours when Cook entered the Guardhouse. Stopping in front of the Desk Sergeant, he said, "Here I am, Sergeant. All set to go."

"Okay, Cook. Here are your orders, your train schedule and a railway warrants. When you arrive in Kimbolten, there should be someone with transportation there to meet you and drive you to the station. Any questions?" The Sergeant the handed Cook the documents.

"What if there is no one there to meet me when I get there?"

"Oh, there will be someone there as they have a copy of your orders and know when your train will arrive."

"How about when I return with the prisoner? Will someone be at the station to pick us up?"

"When you and your prisoner return, check in with Stone baggage. If they don't have a vehicle there at the time, they will call and have one sent down to pick up and bring both of you to the

Guardhouse. You can go on out and get in the jeep now, and the Abbott will take you to the Stone Railroad Station."

Cook examined the documents. The first was Special Orders No. 70 dated 10March 1944; *E X T R A C T*; *Section X:*, Line 14 placed him on TD (Temporary Duty) for two (2) days to AAF Station 117 for the purpose of escorting a Sergeant Stelmeck, held in confinement from that Station to confinement at AAF Station 594, The Repple Depple. It was apparent from the Sergeant's Army serial number, which started with the digit one that he had enlisted in the Army. There were six carbon copies (cc) of the orders held together with a paper clip. The orders had been typed on legal sized, translucent, heavy-duty tissue-like paper using blue-green carbon paper. The copies of the orders had been embossed, three at a time, with an 8th Air Force Seal to make them official. That multiple copies of the orders had been provided was a hold over from times past before there were good and rapid communications between all U.S. Army facilities. A soldier who was traveling under orders may have encountered some Army Units along his way; and to record his passage a copy of his orders would be retained by each of the units. All of the copies were not really needed now, as there existed an Internet of sorts that connected all of the Air Force Stations in England via Teletype[24] over the telephone lines.

The train schedule was listed on a poor quality mimeograph[25] form, which had been prepared using a typewriter and a ruler by

---

[24] The basic Teletype console consisted of a typewriter keyboard and a printer, under a glass window, on a stand under which the operator sat. The keyboard generated electric signals to send text over the Teletype lines to remote printers as well as the console printer. The output of the printer was recorded on continuous accordion folded paper from a box below into a basket behind the console. The Teletype machine was the first I/O device to be used with nascent digital computers because it was there.

[25] To make mimeograph copies, a stencil had to be "cut" using a typewriter and a ruler. A stencil was made by typing and marking on the front sheet of a manifold form to transfer the backing from a porous sheet to the backing sheet. The stencil sheet was mounted on the circumference of a drum. Each rotation of the drum would print one copy of the document.

someone in the Repple Depple. The schedule showed that three train changes were required to reach the final destination. The first change would be in Stoke-on-Trent and was very tight without any layover. The next change of trains occurred in Derby with about a two-hour layover. The last change in Kettering had a two and a half-hour layover, which was more than enough time to have some lunch. The schedule form denoted that the officer in charge was Pvt. M.E. Cook, which taken along with the standard remarks on the form[26], confirmed that the Special Orders had inferred that Private Cook was in command of a detachment of one.

The third document was, in fact, two War Department Railway Warrants. One of the warrants was to be exchanged for a railroad ticket from Stone to Kimbolten. The other warrant was to be used to obtain two tickets from Kimbolton to Stone.

The jeep spun around in front of the station and made a sudden stop. "Here we are," commented Abbott.

Cook knew that they were at the Stone Railroad Station, as he had been there many times while on patrol duty. Nevertheless, he said, "Okay, as soon as I get my bag out of the back of the jeep, you can take off." Just as soon as Cook and his bag were clear of the jeep Abbott let out the clutch and made a sweeping turn around the large empty parking lot. But before leaving the lot he quickly stopped in front of the Stone baggage tent to spend some a little time chatting with the men who were on duty there.

Cook went over to the ticket window and exchanged the first Railway Warrant for a ticket, which he then put in an inside pocket of his jacket, inside of his overcoat. The train would arrive from the

---

[26] Standard Remarks:
1. Person in charge will notify all in his party of their final RAIL destination before they entrain.
2. If additional assistance in required, contact the local R.T.O., either British or American.
3. All personnel are responsible for the loading and transfer of their own baggage.
4. When transfer of trains is necessary, plan to detrain promptly and secure baggage immediately

See Appendix 1 for copies of special orders and travel documents.

South so he to cross over the railroad tracks on the pedestrian bridge. Standing on the open-air platform, he awaited the arrival of the train to Stoke-on-Trent for this first short leg of today's journey.

There he stood, an armed teenager, a foreigner in a foreign land, a buck private charged with an out of the ordinary duty, who reluctantly was setting off on a solo journey to, him at least, an unusual destination. Why had he been singled out for this particular assignment? This trip must have been viewed as being somewhat undesirable assignment; otherwise, some NCO would now be standing on this station platform instead of him. The fact that he was there was a confirmation that his superior officers expected that he would, per his orders, perform this duty by himself without any additional direction. One factor, which doubtlessly contributed to his selection for this task, was that he never really complained very much about, or tried to get out of, an assignment. Six feet tall, well built with a military bearing, his stature, unlike some members of the 1257th, also suited him for this particular duty. He had a purple scar high on his cheek pointing to his left eye, which for this task might make him look somewhat intimidating. While he had better than average intelligence and tended to think before acting or speaking, he was also quite shy and did not have a lot of self-confidence about how he would do on this trip. Filled with trepidation he very much wished that he did not have to take this trip, and given a choice, he would not go at all. But, there was also really no doubt in his mind that he would do his duty and complete this assignment as ordered.

The train coming from the south rumbled to a stop, all the while emanating a hubbub of clattering, clanging, banging and venting steam. The British short journey third class railroad coaches were a series of identical compartments coming and going inside and out. Entrance to the enclosures was through a windowed door centered between two smaller windows. Inside the compartments, with their backs up against the walls, facing upholstered benches spanned between the sides. Space for packages and for baggage was available in racks located high on the walls and under the benches. That the conformation of these compartments was very much like that of a stagecoach shouldn't be surprising, as the first railroad passenger cars were, in fact, the combination of stagecoach enclosures mounted on flat cars.

THAT'S THE WAY THE BALL BOUNCES

When the train had come to a stop, a few doors along the coaches popped open, and some of the passengers exited from the train. Travelers who were waiting to depart from Stone made for one of the open doors, as there would be seating in those enclosures. If they could not find enough seating space in an open compartment, the closed ones would then be checked until open seating was located.

The moment that the train came to a complete stop in Stoke-on-Trent, Cook, with his bag already in hand, quickly exited from the compartment. Then it was a dash up, across and down the overpass to get to the other side of the tracks and speedily find a compartment with an open seat. A guard[27] checked the tickets of all of the passengers who had just boarded the train, and when the last door was closed, the engine with a burst of steam started down the track, to directly be heading east to Derby, Derbyshire.

The British culture of the time was noted for the reserve of the citizens, which in part may have been an artifact of the class system. But it was also a way of coping with the situation like this coach compartment, where strangers must sit side by side for some time. The etiquette of a coach enclosure was the same as that in an elevator (lift) in that you would not normally engage your fellow passengers in conversation, unless of course you knew them. This suited Cook fine, and he was absorbed in watching the English countryside slide past the windows.

About three-quarters of an hour had passed when the train slowed and came to a stop in Uttoxeter. Some of the travelers exited from the train, while at the same time, boarding passengers were finding a seat in one of the compartments. Soon, the train was on its way again, and after about a half hour, it slowed to a stop in the Derby Railroad Station. Cook grabbed his bag and stepped out of the coach onto the platform along with the other detraining passengers, passing by those who were waiting to board.

There would be a little over an hour and a half layover before departure on the next leg of this excursion so he decided to kill

---

[27] In the United States, this person is the Conductor. But, if you stop to think about it, both of these titles are not good descriptive names for a ticket checker.

some time by doing some exploring of this part of the town near the railroad station. This would be his only chance to look around, because on the return trip, he would have a soldier in his custody, even if that layover were long enough. Always wary, with a little bit of vague worry about missing his next train, he did not go far, keeping track of the time and the station location. He was intrigued by the medieval city wall and by the fact that he was only about ten miles from Nottingham. Now, there was a town he knew about from books and movies, but he supposed that Sherwood Forest was long gone. In a slightly related circumstance, he would later marry a descendent of the Scottish Hoods, who would like to think that they were somehow related to Robin.

Heading southeast, the train pulled out of Derby and was soon rushing through the countryside. As it passed through Longeaton, Loughborough, Leicester, and Market Haburough, the train made short stops to allow passengers to get off and to get on. About two hours after leaving Derby, the train stopped in Kettering, Cook picked up his bag and walked out of the compartment onto the platform.

It was already after twelve o'clock, and the next train to Kimbolton was not due for about an hour and a half. He was a little hungry so Cook decided that it was time to have some lunch. He went into the station lunchroom, sat down and ordered a meat pie, mashed potatoes and a cup of tea. The meat pie was made from some ground up mystery meat, more than likely, mutton. The railroad station tea was preferable to the railroad station coffee. After finishing his lunch, there was some time before the train going to Kimbolton would arrive, so he parked himself on a bench, retrieved a book from his bag and passed the time reading.

When the train that he would be taking to Kimbolton pulled into the station, he dropped the book into his bag, got up and headed for the edge of the platform. He found a seat and soon the train was on its way, arriving in Kimbolton[28] slightly over a half an hour later. Walking off of the station platform, he spotted a parked U.S. Army jeep with a driver sitting on it. Cook walked over and addressed the driver, "Hello, I'm Cook. Are you here to pick me up?"

---

[28] The train no longer goes from Kimbolton to Kettering.

"Yup, get in, and I'll take you out to the base." A short time later the jeep stopped at a gate to one of the scattered sites that made up Army Air Force Station 117.

Cook pulled the orders from inside of his coat and showed them to the MP on duty. "I have to go to the guardhouse to make arrangements to take this guy back to the Repple Depple tomorrow morning."

"Okay driver, take Cook over to the guardhouse, and they will take it from there."

When the jeep stopped, Cook got out and extracted his bag, saying, "Thanks for the ride." Cook opened the door and walked into the guardhouse. Walking over to the Desk Sergeant, he handed him the orders and said, "Hello, Sergeant. I'm Cook, here to escort a GI to the 70th Replacement Depot in Stone tomorrow."

"Yes, Cook we have been expecting you. We have a copy of these orders," the Sergeant noted, as he returned orders. The Duty Officer added "We have an empty bunk in the MP barrack where you can sleep tonight. I'll have someone show you where things are, including the mess hall. Corporal Payne, get two clean blankets and a mattress cover from the guardhouse inventory. It will be your responsibility to return them to the guardhouse tomorrow. Cook, you report back here in the morning at zero seven hundred hours. Your train schedule will be ready, and you can take custody of your prisoner. There will be a vehicle here to take both of you to the train station."

"Yes sir. I'll be here ready to pick up my prisoner at zero seven hundred hours."

"Okay, Let's go. I've got your bedding, and I'll show you where to bunk", Payne added, leading the way as the two exited from the guardhouse.

As he made up his bunk, Cook and the off duty MPs who were in the barrack engage him in good-natured but vulgar banter and amusing insults. It was chow time, so they all took off for the mess hall, where he was provided with some eating utensils that were available for casual soldiers.

This was one of ten dispersed sites that made up Army Air Force Station 117, and there was nothing that indicated just where site number one, the actual airfield, was located. Cook had no knowledge or real interest as to just which Eight Air Force unit was stationed here. It would have meant little to him at the time, and it was probably classified information in any case, but he still would have liked to haven seen some bombers. In fact, Kimbolton, Northamptonshire, AAF STA 117 was the location of the Eight Air Force First Air Division, 379th Heavy Bombardment Group consisting of Bomber Squadrons 524th through 527th.[29] At the time, the 379th Aircraft Compliment was 48 B-17G Fortress bombers.[30]

At about zero seven hundred hours, Cook was back at the guardhouse, ready to take custody of the soldier whom he was to convey as a prisoner to Stone. The MP bassard was now located on the upper left sleeve of his overcoat, and his white web belt had been adjusted to sling the sidearm on the right hip. Coming to rest in front of the Desk Sergeant, he said, "Sergeant, I'm here ready to take charge of the prisoner. Do you have the railroad schedule ready?"

"Yes, Cook, this is your train schedule, and I have here a receipt for custody of your prisoner for you to sign. It will be about forty-five minutes before someone takes the two of you to the station. Your man is almost ready to go, and as soon as he is, we'll bring him out of the cell so that you can get acquainted with him."

Cook bent down and signed the receipt, taking a copy and then he picked up the train schedule, putting both in an inside pocket. "Okay. Does he have any baggage?"

"No. Everything he has now is what he is wearing, except for his personal property which is in this envelope" the Sergeant said as he handed it to Cook, who took it and put it in his bag.

The use of restraints on a soldier in custody was not even a consideration for such a minor violation as being AWOL. In any case, Sergeant Stelmeck was very amiable and cooperative, and it was evident that he did not intend to cause any problems. Private Cook was now in command of a detachment of two. The guardhouse door opened and a GI came in, "Sergeant, I'm here to pick up two men and take them to the railroad station."

"Yes, they are ready to leave. Cook, you and your man can go on out and get in the vehicle, and the driver will take you to the station."

---

[29] See endnote 7 for sources of information on the Kimbolton AAF Station and the units that were located there.

[30] See endnote 8 for the B-17 Fortress Bomber parameters.

"Okay, Sergeant. We are on our way, and thanks for everything."

"Let's go" the driver said as he headed out of the doorway.

"Okay, Sergeant. You follow the driver, and I'll be right behind you," commented Cook, closing the door as he exited the guardhouse. The driver climbed up into the weapons carrier and started the engine. "You get in first and sit in the middle, Sergeant," directed Cook as he pulled himself up on to the seat. "Okay, let's go."

The weapons carrier came to a stop in front of the railroad station. Cook swung out to the ground, turned around and watched as the Sergeant dismounted. "Thanks for the ride, guy." "Okay, Sergeant, let's go over to the window here so I can get our tickets." Cook removed the packet of orders from inside his coat and pulled out the second War Department Railroad Warrant, which he then exchanged for two tickets to Stone. "Now, we'll go stand on the platform and wait for the train going to Kettering, which should be here soon."

It was early Sunday morning, and there were few travelers standing on the platform. When the train arrived, they easily found an empty compartment, which they entered, sitting down on opposite sides. Cook placed his bag on the seat by his side. There was the usual talk of where in the U.S. they had come from and about events of the war. Just why the Stelmeck had been confined was not a subject to be addressed. For most of the time, they just watched the English Countryside slide past the windows. After about a half an hour, the train came to a stop in Kettering. They left the compartment, found a bench and sat down to wait for the train to Derby.

There were more travelers walking about and arranging themselves on the Kettering platform waiting for the train to Derby, which finally arrived after about an hour and a half. "Okay, Sergeant, let's go stand on the platform and see if we can find an empty compartment, and if not, try to find one without very many people in it." After checking a few compartments, they found one that had some open seats. Following his prisoner into the compartment, Cook put his bag into the luggage rack and sat down on the left side of him. Now, they sat in silence watching the green countryside slide by and the passengers leaving and entering the compartment at the stops along the way.

When the train stopped in Darby, after about a two-hour trip, Cook grabbed his bag and followed his prisoner out onto the platform. It was more crowded with travelers getting off and on the train than it had been in the Kettering Station. It was after thirteen hundred hours, and they had about an hour and a quarter before their next train departed. Cook said, "Let's go to the station lunchroom, and I'll buy us some chow." After finishing lunch, they went and stood on the platform awaiting the arrival of their train to Stoke-on-Trent. Many more travelers now crowded the platform, which in turn, made finding a compartment with two abutting seats even more difficult. Once seated and underway, it was the same routine of watching the countryside and the passengers getting off and on when the train stopped during this hour and a quarter segment of the trip.

When the train arrived in Stoke-on-Trent, they had only about a half an hour before they would board the train for their final short ride. They quickly crossed over the tracks on the pedestrian bridge and took up a station on the platform. When the train came to a stop, they hurriedly located a compartment with two empty seats and sat down. A short time later the train was on its way south to Stone. Looking out of the windows, Cook now saw countryside that he was familiar with. After about twenty minutes, they were in Stone where they exited the train. Cook felt a little excitement now that he had finally returned to the very spot where he had started this railroad excursion so very long ago yesterday morning.

"Okay, let's go on down to Stone baggage, it's that circus tent on the right, and get a ride out to Howard Hall." Walking down the sidewalk, they came to a gateway and turned into the entrance of the tent. "I was told that you would arrange for a ride for me and my prisoner out to Howard Hall."

"Yeah, we have been expecting you. Get up in the front of the truck, and I'll haul you out to the Repple Depple."

A quarter of an hour later, the truck stopped to be checked in by Andriko who was on duty with Colcord at the Howard Hall gate. "Okay, we'll get off right here. Thank you for the ride," Cook said as he swung down to the pavement. "Come on Sergeant, get down, and we'll go into the guardhouse." The truck continued on up the road to a spot where it could be turned around, and then headed back through the gate on its way back to Stone baggage.

"Hi, Cookie. I see you're back," noted Andriko

"Yeah, and I'm glad to be back," and with that, Cook directed his prisoner through the guardhouse door. It was after seventeen hundred hours when they arrived in front of the Desk Sergeant. "Hello, Sergeant, this is Sergeant Selmeck the GI who I was ordered to escort back here. You can book him into the cell now." Reaching into his bag, Cook pulled out an envelope, "This is his personal property."

"Okay, Cook. You are relieved of your prisoner," the Desk Sergeant replied as he started to log Stelmeck into the guardhouse cell.

"Say, Sergeant, here are the receipts for my lunch yesterday and for our lunches today. What do I do with them?"

"You leave them with me, and I'll find out how to process them. Cook you don't have any duty for the next two days, so if you want to you can take off on a pass in the morning." Bane, who was also on duty unlocked the cell door, and then told Sergeant Selmack to go on in and to make himself at home. That would be the last that Cook would see or hear anything about his erstwhile prisoner, who would in a short time be subjected to a disciplinary hearing. The hearing would result in such punishment as reduction in rank, forfeiture of pay, restriction to station and extra duty, but probably would not exact any additional confinement.

Very happy that he was now back from his little excursion, Cook had also gained some self-confidence and self-assurance, but was not really aware of that at the time. It was after seventeen thirty hours, and Cook thought, "I'm going to my room, put away my gear, change some clothes and get some chow, because I'm hungry." He was so hungry, in fact, that he really did not care that the mess hall was serving the traditional Sunday supper of cold cuts.[31]

But still there was the question: In just what manner had this mission advanced the progress of the war against the Axis Forces?

---

[31] The military vernacular for cold cuts has the initials "HC" with the first word being "horse."

Chapter 28

Tales from the Barrack-Cleaning Detail

While everyone who lived in the barrack was responsible for keeping their personal quarters (their shared room) clean and in order, the rest of the barrack—the halls, the Common Room, the Utility Rooms and the Latrines—were the responsibility of the cleaning detail. The cleaning detail was mustered from the lower ranking members of the Company, who received their assignment from the higher-ranking members.

One morning while on duty as a member of the cleaning detail, it was my task to mop floors. As I plied the mop across the variegated dark brown asphalt tile covered floors, I was, as soldiers are disposed to do, engaged in a philosophic discourse with one of my comrades, The Bad Penny. I refer to him here as The Bad Penny based on the confluence of two elements. First, in real life, his nickname had apparently been "Lucky Penny," which name he had someone emblazon, complete with embellishments, on the back of his field jacket. This was in emulation of the fashion of airmen having artwork, including names and slogans, painted on the back of their leather flight jackets. The other element stemmed from the fact that one time or another, he had borrowed money from a number of men in the 1257th, myself included,[32] and of course was unable to it pay back; not that he really ever intended to pay it back. But then he did have some good and some fantastic stories of how he would soon have the money that he owed to you.

---

[32] In an old record book I found a list of my loans to various members of the company. This loan amounted to ten Pounds Sterling, which at the exchange rate fixed for the American Forces was equal to forty Dollars.

THAT'S THE WAY THE BALL BOUNCES

As the philosophic discussions continued, the subject turned to how in the development of events, after the initiation of some action, things could soon be completely out of control. I said, "Yes, when you throw a ball, after the first bounce, you no longer have any control of where the ball will now end up. What happens is what happens. It's just *the way the ball bounces.*" We both thought that this was a very good expression and would use it in any situation where things could not be kept under control. Soon in the 1257th *that's the way the ball bounces,* became a catch phrase for any situation that could not be controlled. I more or less forgot about the expression after I was out of the Army, and that was the end of it, so I thought.

During the Korean War, three expressions for fate became part of the popular idiom. One was *that's the way the ball bounces,* and who knows if this catch phrase had spread to common usage like ripples on a pond from that chance observation on that morning in the MP barrack at Howard Hall. Now, it is quite possible that someone else, at some other time not long after, had come upon this very same phrase. Did this expression propagate from a second chance observation? What are the odds?

I was looking in (Rodale) *The Synonym Finder* for a synonym meaning "fate" (it is also for "chance") and there it was *that's the way the ball bounces.* [33] This was somewhat curious as the genesis of this phrase was to express lack of control, but no matter, I still felt a little flush of pride. However, there was an irony because I was mopping when *that's the way the ball bounces* was first expressed, and *that's the way the mop flops* was also one of the catch phrases for "fate" which was current during the time of the Korean War. But, *that's the way the mop flops* did not make the cut, and *that's the way the cookie crumbles*[34] which did.

---

[33] (Ammer} *The American Heritage Dictionary of Idioms* also defines this phrase and indicates that the origin as colloquial: mid-1900s. (Which is consistent as it was 1944.)

[34] This catch phrase has largely fallen out of use, as there is little social need for two different catch phrases that express the same idea.

Chapter 29

Guarding the Gate-Always on the Watch

The vehicles that passed through the gates of both Howard and Duncan Halls were viewed in two categories: those that we did not stop, and those that we did stop. Foremost in the first category was the black, sporty Jaguar with amber headlights, driven by the 70th Replacement Depot's commanding officer, Colonel Rader, which passed through the gates however unimpeded. Also exempt from stopping were officers in staff cars, officers whom we recognized in other vehicles such as jeeps, and of course all vehicles driven by MPs.

The second category of vehicles was primarily the trucks that were hauling supplies and personnel. All of the vehicles within this group had to come to a stop whenever it passed through a gate to be logged in, or to be logged out, by an MP on duty at the gate. The U.S. Army vehicle number, along with information listed on the trip dispatch document such as the driver's name, station number and the time of day were entered on a loose log form on a clipboard. Rain, which was not uncommon, added a level of difficulty to making entries on log sheets, a reduction in vigilance and sometimes errors. In any case, the logging of vehicles into and out of the station was hasty, and very little consideration was given to the actual authenticity of the trip dispatch documents. We just had to assume that the documents were okay because we didn't have any way to check them out in any case, which was probably typical for all other stations. And so there never was an absolute certainty that all of the trucks passing through all U.S. Army station gates were on official business. (So much for most "security" which at its best is an elusion based on mutual assent, even in the totalitarian state. The saving grace is that usually those who would breach the "security" do not have any more imagination than those who are the **security professionals.**)

THAT'S THE WAY THE BALL BOUNCES

The insatiable Draft Boards, required to meet quotas for warm male bodies, uncritically sucked men in from all walks of life. Included in this maelstrom[35] of men being conscripted for war were both practicing professional criminals and would be amateurs, who would rise to the occasion whenever a golden opportunity for malfeasance was perceived. Now, the fact that one was a legitimate or a prospective criminal does not preclude him from being a patriotic and sincerely serving his country. For instance, one of my comrades in the 1257th had been, in real life, a professional moon-shiner, and a very good one to hear him talk about it. But regardless of his trade, his performance as a soldier and as an MP was as exemplar as that of any other member of the Company. Also, to his credit, he did not engage in any "moon-shining" while he was doing his duty as a member of the United States Army Air Force.

There was a certain and present level of confusion attendant to the continual and ever expanding influx of men and material in Britain as the preparation for the invasion of Europe accelerated. This environment provided a capital opportunity for both the amateur and professional Army criminals to prosper. It was a classic case of supply and demand; the U.S. Army had the supply of food, fuel, and whatever, and the British black market provided the demand. AWOL soldiers with AWOL trucks were in cahoots with GIs, with access to supplies, and who also made up false paperwork so that they could cater to the black market.

While the Army C.I.D., the Criminal Investigation Division, the U.S. Army's own detective agency, had its work cut out for it with all of the run of the mill Army crime, it was very difficult to run down and apprehend the AWOL trucks and drivers. There were so many AWOL trucks and drivers running about who were engaged in supplying the black market. So in order to catch them unawares, at random intervals all of the regular U.S. Army vehicular traffic and all of the passes would be suspended for twenty-four hours. MPs would be sent on patrol searching for errant GIs strolling about; GIs who, unless they actually had leave papers or orders, would be AWOL and thus be detained. To catch the errant trucks and drivers, rolling on their merry way down the highway, a series of roadblocks were established on all of the major roads.

---

[35] See endnote 9.

One night on a local road, during one of these re-occurring roadblocks, the two MPs on duty observed a V-1 Buzz Bomb[36] flying over at a low altitude. This must have been a very unusual occurrence, because there had never been any other reports, which I was aware, of a V-1 flying so far north in to the midlands of England. However, to the MPs manning the checkpoint their duty was clear. So, they drew their side arms and sent skyward a barrage of .45 caliber antiaircraft fire. While this assault was at best futile, but they could not be reproved for their action as this would be the only opportunity for any soldier in the 1257th to actually engage the enemy.

The fact that this particular V-1 Buzz Bomb had flown this far north was probably the combination of a favorable tail wind and the failure of the guidance systems mechanical counter. The counter would open a switch after completing a preset count down that turned off the fuel to the pulse jet engine, at which time the missile would go into a steep dive, exploding on impact. Now there was one instance when the V-1 had flown even further north. On Christmas Eve 1944 the Germans had air launched forty-nine missiles from bombers flying over the North Sea. But, only one V-1, of the thirty-five that impacted in England, actually hit the intended target, Manchester. However, our errant Buzz Bomb did not fly by on Christmas Eve as that would not be a very effective time to catch AWOL trucks, as few could be expected to be out and about. Also, it would have been highly detrimental to morale to have a stand down (a term which was not used at the time) at this very special time of the year.

---

[36] See endnote 10 for parameters of the V-1 Buzz Bomb.

Chapter 30

Singular Service-Point Control

On the way towards Stone, the country road that ran through Yarnfield quickly dropped down a short grade as it passed through a crowded stand of trees and terminated at Highway A34. To go any place east, north or south in the English midlands from the Repple Depple, the most direct way was via A34. Turning onto A34 north or south, you would soon come to a turn off onto a country road going to Stone, with a left turn providing the shortest way. There was ever the probability that some GI truck driver, driving on the "wrong" side of the road (i.e. The driver's side of the vehicle was to the outer side of the road, while the passenger side was to the middle of the road.), would zoom down the grade and, with much bravado, make the blind turn onto A34 while hardly slowing down. To forestall such a maneuver, and thus prevent an almost accident involving an U.S. Army vehicle, an MP was posted in the center of the road just where it ended at the highway. There was an MP on Point Control duty there from early morning, until the last man on the post was picked up when the MP jeep returned from Stone with the evening MP patrol. In addition to controlling the U.S. Army vehicular traffic, the MP on the Point also had to provide direction to anyone trying to locate one of the local elements of the far-flung Repple Depple. An important part of the MP's duty was to assure the safe crossing of A34 by the GIs, who out on a pass, were walking to and from Stone. There was a clear and ever present danger that soldiers fresh from the states would look the wrong way at this blind corner before crossing the highway.

In any military organization, there will always be an enterprising few that will engage in making a buck by offering something to their fellow service members. These members are ready to sell whatever is not readily available such as various soft drinks, candy, food-stuffs and reading material. Some will even go farther to provide services such as transportation into town. At the Repple Depple, this last service was provided by one of the permanent party NCOs who had set himself up in the bicycle rental business. He had acquired a fleet of bicycles that he would rent out to some of casual GIs on a pass, a GI who would rather ride than walk or take a chance on a bus to Stone. The Sergeant's bike rental operation, while unapproved, was not clandestine, but had been overlooked, maybe due to lack of knowledge, by whomever should have stopped his enterprise.

The MP on duty at the Point one evening reported that two GIs riding double on a bicycle came flying down the grade. He said that just before they smashed into a tree, one of them had yelled out, "This is more fun than a barrel of monkeys." This would be the last words either one of them would hear or say. The next day, the Sergeant was out of the bike rental business.

Point Control at the Yarnfield road to A34 connection was not one of my regular assignments. In fact, I was only posted there one time when, I suppose, the MP who was scheduled for duty there was for some reason not available. It was the morning shift and was pretty much routine, that is if you don't mind standing in the middle of the road with GIs driving by on the "wrong" side. There was one very interesting event, not on the road but in the sky where I saw a very large flight, maybe an Air Division, of B-24 Liberator Bombers[37] in combat formation passing over Stone. I wondered if they were formed up for a mission. This was the only large formation of heavy bombers that I ever saw, and I was not aware that the formations ever came so far west. I had never seen any bombers in formation at Howard Hall, which was located only a few miles farther west. To me, it was an incredible and exciting event seeing that formation of heavy bombers.

---

[37] See endnote 11 for parameters of the B-24 Liberator Bomber.

CHAPTER 31

Guarding the Gate-A Woman of Mystery

It was a bright and sunny day, the best of British summer afternoons. We were on duty at the Duncan Hall entrance gate on this very nice day. My partner, whose name I cannot remember, was a later replacement member of the MP Company. Many of my comrades did not like to be assigned to duty with him for a very strange reason. He had grown up in an extremely rural area, worked on the family farm and never attended school. He was in fact illiterate when he was drafted and the U.S. Army had placed him in a training program where he was taught to read and write. When he had mastered the minimum Army requirements, the Army in its perverse logic assigned this "big, dumb guy" to the military police. He was, of course, not unintelligent, he was just uneducated, but there are so many people who cannot discern the difference.

The traffic entering Duncan Hall this afternoon was normal when unexpectedly, two women, in their mid-twenties I think, came walking through the gateway and came up to us in the gatehouse and stopped. Now this was very unusual. While there was always some foot traffic to and from Howard Hall, visitors to the station never came on foot, and were almost never civilians. We had not noticed them on the road that ran past Duncan Hall, and did not have any idea how they had made there way to the gate. Maybe they had walked back from the bus stop, or maybe they had walked from a car that they had parked in Yarnfield.

One of the young women, the leader, was a beautiful, svelte, strawberry blonde, and the other young woman, while very attractive, was not in the same league as her companion. Speaking with an English accent, the charming beauty told us that she was seeking a fighter pilot, whom she had known in Rangoon, who was a very good friend, and whom she believed was here at the Replacement Depot at this very time. She wanted us to let them go into the station to try to locate her friend. I told her that we could not let them go into the station. The she wanted me to call and see if I could find out if her friend was here at this station. I said that I could not do that, but she could go over to the guardhouse, at Howard Hall, and see if they would help her. The truth was that I did not know whom to call to see if I could find out any such information, but it did not matter, as I would not have called in any case.

After some small talk, the two young women when on their way, walking out of the gate and turning toward Howard Hall and Yarnfield. If they then went over to the guardhouse, I do not know. If they had gone there, I think that I would have heard about their visit. Someone as good looking as this lady showing up at the guardhouse would be sure to have generated some sort of speculation, and I think that I would have heard the tale.

The beauties story, while very romantic, the stuff of Hollywood, was somewhat improbable. The idea that you might find someone for whom you had been diligently searching at the Repple Depple was strange. In either case, if he had just arrived from, or was just about to return to the United States, it would only take a very few days to process the necessary paperwork and arrange transportation. Why would she believe that her friend would be here at the Repple Depple at all and that if he were here why would it also be on just this particular day. Had she searched the whole world over trying to find him and had just now obtained information, and how had she obtained it, that he had just arrived from or was about to return to the United States, and she wanted to see him just now in any case. This charming one's story of searching after a lost-love, would have made some sort of sense if she had been looking for him at an airbase in East Anglia, but it did not jibe with the operation of the Repple Depple. The question is was she in fact working for the U.S. Army checking the security of the various Air Force stations? This may have been the case, but I will never really know the truth. It was all very mysterious.

DUNCAN HALL ENTRANCE (NO.2) GATE HOUSE

Here I stand, ready to repel any one who would attempt to breach our security, so watch out.

Chapter 32

Tales from the Barrack-Packages from Home

As soon as the 1257th was in place as a unit of the Repple Depple permanent party, our mail delivery was regular. Letters, packages and magazines arrived frequently from the United States without substantial in transient delay. While the mail delivery to the MP Barrack was regular and frequent, the actual amount of mail received varied with the individual. For one year, I was receiving a subscription to Esquire magazine that was a gift from my buddy, Vernon. The quantity of mail that any one received was most likely a function of the particular antecedents and family relationships of each soldier. Most couples, married and other wise, wrote to each other as often as every day but did not always put their letters in the mail every day. I made it a point to write home to my mother at least once a week without fail. All of our outgoing mail was reviewed and censored by one of our company officers, which depending on their workload might sometimes result in a small delay in the journey of a letter home. On the front cover of the letter, or in a special box in the case of in V-mail, all outgoing mail was imprinted with the Army Examiner's Stamp and signed by the Actual Censor.

The regular mail was so fast and so reliable that V-mail was not generally used. The Army had encouraged V-mail as a method of sending and delivering mail quickly under adverse circumstances. Also, V-mail did not require much cargo space and could be transported by aircraft. The letters and sometimes greeting cards were composed on a special form, which was stamped with a serial number and then photographed onto a roll of film. The quickly developed negative was then sent quickly to its destination where a photocopy of each letter was printed using a special V-mail processing machine. The reproduced copies were then sent via the regular local mail

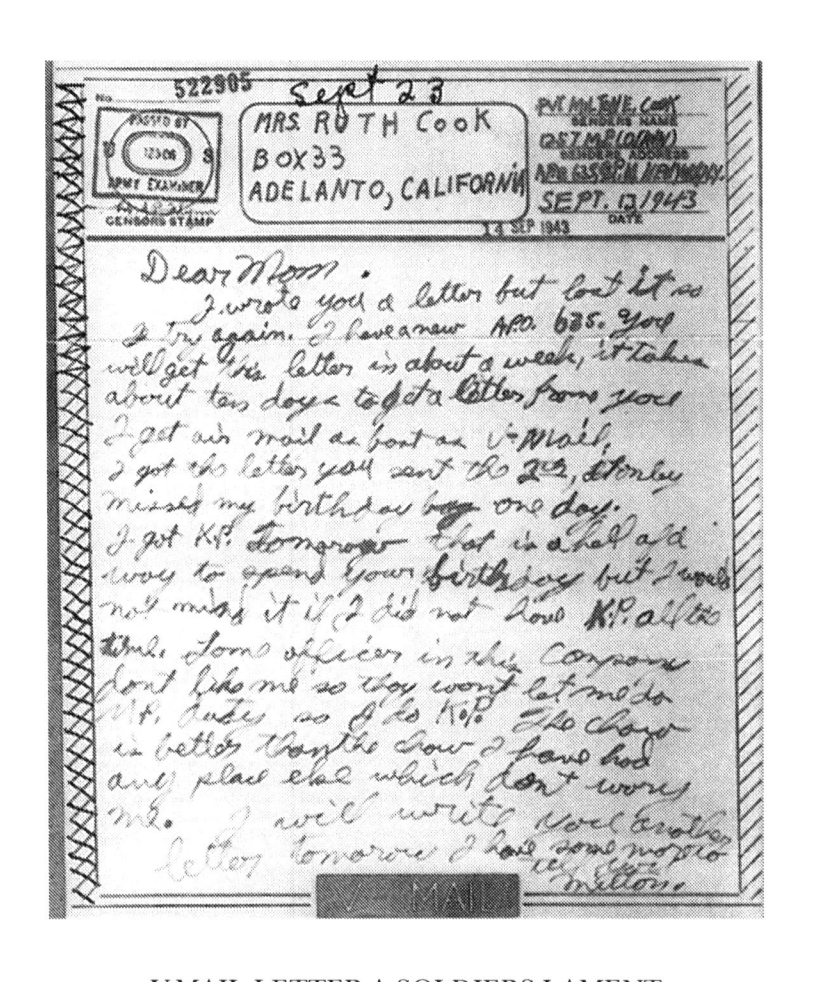

V-MAIL LETTER-A SOLDIERS LAMENT

> At this point in time the 1257th had been at the Repple Depple for only twenty-seven days and were still billeted in a Beatty Hall barrack. Most probably there was some policy in place that all arriving units had to provide men for cleaning details and for KP duty. I had been on a cleaning detail in the Beatty Hall commissioned officers lounge before being assigned to KP duty. In our case this was a very short-sighted policy as it both diverted men from what was to become and intensive duty and it was an assignment which would also some what compromise their effectiveness as MPs. At any rate when our small detachment returned from TDY at Chorley the 1257th had relocated to a Howard Hall barrack, that was isolated from the rest of the Repple Depple permanent party and now there were no extraneous assignments. (This particular V-Mail letter survived because it was the very first one that my mother received. And so being the first of a kind my sister had pasted it in to her war time scrap book.)

system to their final destinations. Before being photographed, the V-mail letters were sorted into groups having a common destination such as city and state when going to the U.S. and the specific military unit when going to one of the theaters of operation. Much like a postcard, a V-mail letter could be read by anyone handling it along the line, which probably did not make them a more popular method of sending letters. The original V-mail form was printed in a light red on an 8 ½ by 11 inch sheet of bond paper. When the missive to home was completed, the sheet was folded from the top and the bottom to meet at the center of the page, and then it was folded once again at right angles to form a 4 by 5 ½ inch wide envelope/letter. The serialized copy of the V-mail letter was about half the size of the original and was folded to show the name and address of the recipient in the oval window of a 3 3/4 by 4 5/8 inch brown paper envelope.

I did comment earlier about receiving a hand-knit OD wool sweater vest from my mother, but for all of us, the real highlight of mail call was to receive a package from home. Typically, the package would contain something good to eat, more than likely cookies, which we would share with everyone in our own platoon. My mother would frequently send me a package containing her homemade cookies. The stacks of cookies were wrapped in waxed paper, and then these packs of cookies were packed in popped popcorn. The popcorn had not been seasoned and was also somewhat stale, but it was still edible, if of course you were not very picky. Some might believe that those small pellets of plastic foam packing material are called popcorn because they somewhat resemble popcorn. But, no, they are so called because they have replaced real popped popcorn, which is after-all also a very good packing material, and of course is biodegradable.

The popcorn packing material gave me an idea, so in a letter, I asked my mother to send me a bag of popcorn kernels. Using my mess kit and some butter from the mess hall I made small batches of popcorn on the hotplate, which we had in our room for just such applications. So, we could have fresh popcorn in our room when we had some free time and decided we would like to have some.

A really uncommon treat in a package from home was the oranges, which Collins occasionally received. His uncle was an engineer who worked with a citrus packing plant, and who was working on the development of the machinery and process to wash and wax citrus fruit.[38] Collins' Uncle would occasionally send him a box of oranges as an experiment to find out how well the waxed fruit was preserved and how well it survived the shipment. The fruit was very tasty, appreciated by all, and we were most probably the only group of GI's in the ETO who regularly enjoyed specially delivered fresh California navel oranges.

---

[38] He was probably a member of the University of California, Riverside citrus station staff.

Chapter 33

Off Duty-Passing It Around

Some of my platoon mates, those with whom I regularly partnered on patrol duty and on gate duty also became my buddies. A buddy is a guy with whom you bond and become friends, as a result of shared duty and adversity, and enjoy some common experiences. Szucs[39] was just such a buddy, as we had walked many a patrol together and had stood many a shift of late duty together. So it was only natural that frequently when we were off duty at the same time, we would head someplace on a pass together. The one place where we would regularly end up was the town of Hanley, where for two-en-six (50 cents), the American Red Cross Club would provide a bunk and breakfast. In fact, the Red Cross Club was the only place in that part of England with facilities in which a GI easily could stay at night. Hanley attracted U.S. soldiers from stations all over the midlands and strangely there were no MPs, except of course for those of us who were here off duty. The accommodations in the club we a common latrine and double bunks with bedding. Except for not being in an actual Army barrack, all of the bunks crowded together side-by-side in large rooms was more like normal Army life than the facilities in which we were quartered. In addition to the regular canteen fare of donuts, coffee and cokes, the Club served an English breakfast for those who stayed overnight. The breakfasts served were typically British with such items on the menu as kippered herring, bangers

---

[39] The "S" is silent and the Brits were forever asking him how he spelled his name. He had learned to reply "S," "Zed," "U," "C," "S." The British name for the letter Z is Zed, which make the name of this letter somewhat of an alphabetic oddity.

and lots of toast and jam. The American Red Cross obtained as much as possible of their supplies locally. Almost all of the workers in the Red Cross Clubs were local residents.

HANLEY RED CROSS CLUB

Passing the time during the day in Hanley, Szucs and I might spend some time visiting some of the variety of shops found along the side streets. In one of these stores, Szucs had discovered sometime in the past that the shopkeeper had at one time been involved in gold mining. So when we stopped in this shop, he introduced me to the proprietor telling him that I had also been involved in gold mining. This meeting resulted in a long conversation about our individual experiences and common knowledge of mining for gold. Sometimes we would head down the grade towards Stoke-on-Trent to visit Hanley Park where we would just wander around and some times take a few photographs. Szucs had acquired a used reflex camera in one of the shops, and as film was readily available in the PX, he was taking pictures wherever we were. He developed the film and made some prints in the station hobby shop darkroom.

Sometimes, we would take a bus trip through the countryside to some local spot, just acting like tourists. One day he said, "Let's go out into the country where I want to see someone." So, we were off on another bus trip, but this time to a mystery location. We were rolling through farm country when he said, "We get off at the next stop." There we were in the middle of nowhere with hedgerows lining both sides of the road. We headed back down the road and soon came to and turned onto a country lane that we had passed on the bus just a short time before. Walking down the lane, we eventually arrived at a group of farm buildings. It turned out that we were here to visit a girl who lived on this farm. Apparently, Szucs had made some sort of a date to come out here in the country to see her. The farmhouse was large and stylishly decorated, obviously an upscale prospers farm. After spending some time looking over the farm we set off making our way back down the lane and road to the bus stop, where we eventually caught a bus that was going back into town. I do not know if something was supposed to happen at the farm, but nothing did. Maybe that is why he had me come along on this strange excursion into the English countryside.

On some other occasions, we would be off on a train trip to someplace usually to some nearby town or place. One day, we went as far south as Birmingham just to see the city; although this trip was in fact well beyond the limits of our class "B" passes. While there in the Red Cross Club, I saw the only American war correspondent that I would ever see in the ETO. Twice we went down to London on a two-day pass. A special pass was required to visit London, in order to control the number of GIs who were there at any one time, and to make it more difficult for GIs to go AWOL there. This would be the only two times that I stayed at an American Red Cross Club in London. On all of the other times that I was in London, I was on duty and so was quartered in a U.S. Army facility. We did the usual sightseeing, wandering around with Szucs occasionally taking a photograph. In the morning, we went to Buckingham Palace to watch the changing of the guard. We arrived there early enough to stakeout a good location on the upper level of the Queen Victoria Memorial to observe the ceremony. The guard units were dressed in their wartime Khaki uniforms and not the more familiar dress uniforms. Even so, there was a good-sized crowd that was watching this famous routine.

THAT'S THE WAY THE BALL BOUNCES

TWO DAY PASS TO LONDON

> A late issue standard 48-hour pass which was required to travel beyond the area that was allowed by a class "B" pass. The cancellation clause appears to have a catch 22. Any soldier who was ordered to report directly back to his unit might be considered to be loitering while waiting for public transportation. Also, at the very bottom is one more shot in Uncle Sam's war on VD.

VIEW OF STREET FROM RED CROSS CLUB WINDOW

CHANGING OF THE GUARD—OCT 1945

Frequently, we would make the trip to Hanley to meet our girlfriends, sometimes together and some times solo. Szucs had, at some time when off to Hanley by himself, met and starting dating Barbara Clews. As it happened Barbara had a friend Lillian Shlinger, and so it was arranged that I would meet her on a blind date. Both girls lived in Stoke-on-Trent. Sometimes, we would be going someplace with these two young ladies on a double date and sometimes, I would just meet Lillian, depending on our duty assignments. There were two first-rate cinemas, almost across the street from each other, which showed both American and British films: both recent films and reissues of some of the classics such as *Snow White and the Seven Dwarfs* and *Gone With the Wind*. One time when we were watching a movie, we saw a shot of an airplane flying over some southwestern desert hills covered with scattered sagebrush. I told Lillian that the desert scene looked a lot like the countryside of where I lived, and you know, she was not the least bit impressed.

Lillian had told me that she had a boyfriend who was a soldier in the British Army and who was away from England involved in a war in some other part of the world. This was an example of one of the great incongruities of the wars. Here, there were a multitude of British Empire service personnel deployed around the world engaged in various aspects of the bestrewn conflicts, and at the same time, here in England, there were teeming masses of American attacking Germany and marshalling for the soon to be invasion of Europe. Somehow, this all seemed to me to be a peculiar apportionment of manpower, but of course, there was no rational solution for this curious conundrum. As it was each country had to take care of its very own particular vested interests in its own peculiar way.

Food and eating are always an important part of dating. One of the repasts that we particularly enjoyed was that "renown" finger food, fish-en-chips. Greasy battered cod and greasy sticks of potatoes deep-fried in real, honest-to-goodness tallow. Dumped onto fresh, honest-to-goodness newspapers, and then liberally doused with real malt vinegar to cut the grease, this was a truly delightful meal served in a humble fish-en-chips shop. I had been told that there was, although I never saw any, an English version of the American potato chip that was called potato crisps, which is of course a very logical name.

On an overcast, summer Sunday afternoon, Szucs, Barbara, Lillian and I took a bus ride to a local park, where boats could be rented and rowed around a small lake. There was no one else on the lake, but we rented two boats and off we went on our merry way. Now, Szucs was from Michigan, and I suppose he had some experience handling rowboats. I, on the other hand, was a complete novice. The only boating in the part of the Mojave Desert in which I lived was on the lakes in the nearby mountains, most of which I had never seen at that time. The boats had a very shallow draft and when coupled with my complete lack of experience in small boat handling, I managed to ship quite a bit of water into the boat during one inept maneuver. As my half-Wellington boots were already wet, I removed one and used it to bail out the boat. All was well, but the end was a little wet.

On some nights, the four of us would enjoy the evening in a pub having a few pints of beer. On one such occasion, we were enjoying ourselves with witty and clever repartee, when at the ten o'clock closing time, a local who was sitting at a nearby table stopped by our table and told us that we had really been very silly. Well, so much for British reserve and tact. We were just a group of teenagers trying to enjoy ourselves in a "crazy world," didn't he know that there was a war on?

Chapter 34

Guarding the Gate-A Rueful Account

Late one afternoon, on duty at the Howard Hall Gate, my main activity was checking passes of the soldiers as they streamed out headed for the local spots with pubs. The first wave of the exiting GIs had pushed on their way, but as always there were a few who were in no hurry to head out from the station. One of the men lagging behind, who I knew in passing, was a member of the permanent party who were quartered in Beatty Hall. I checked his pass, but instead of going on his way, he stopped to talk to me. Just why he had selected me to hear his disclosure, I never understood. Maybe he believed that I was circumspect and would not blab what he was about to divulge. Now, he might have gone to talk with the Station Chaplain[40], but it was probably much easier for him just to spill it to me and get it off his chest.

He said that he had a wife and several children back home in the states. In one of his letters home, he had asked his wife to send him a camera. He had received the camera, and then in a drunken rage, he had smashed it to pieces by slamming against the wall in his room. He told me "I'm so very ashamed of myself for doing such a terrible thing, and I don't know what or how to tell my wife."

I said to him, "Yes that was a very bad thing to do." But, I did not ask him just what it was that had caused him to demolish the

---

[40] "Go see the Chaplain and have him punch your T.S. card" was a common expression in the Army when someone was complaining unreasonably about something that nothing could be done about. T.S. are the letters which stood for tough stuff, where stuff is the same stuff referred to in footnote number 19.

camera. The reason was because I knew well that the lingering low level of pervasive distress that we all experienced was the probable cause of his behavior. Our living conditions were good compared with those of most of the Army in the ETO. Still there was always an undercurrent of adversity due to being in the Army, to being long and far away from home and to not knowing if or when we would ever be able to go home.

I think that most of us suffered from homesickness from time to time, some severely and some not so very much. But in any case, it was not a subject that we could or would ever discuss. Each and everyone had to cope in their own way as best they could. To me, it was worse in the still late hours of the night when I was in my bunk trying to go to sleep, thinking about home and how it was when I was last there. It was strange, but I had my mother send me a set of eight snapshots that I had taken during the short time after I graduated from high school, while I was in the mountains with my father and brothers. This set of pictures somehow provided some respite and escape, for at least a short while, to a happier time. All in all, it was a very distressing experience—not only the being away from home and family for such a long time, but also the not knowing how long it would be before I would ever return home. At the time, the duration of the stay in England seemed to be forever without end.

CHAPTER 35

Guarding the Gate-Reunions

As the American Forces convened in England, the chances increased that a sibling or some other family member, who was in the U.S. military, would show up somewhere on the island. One day, Sergeant Foiles received a message, sent APO to APO (Army Post Office), from his brother stating that he was now in England. Foiles was eager to go and visit his brother, so he found out the location of his unit, and he asked me to accompany him on the journey there. Special passes were required to make the trip to his brother's station. Most of the journey was by train, but I do not remember just where our destination was in southern England.

His brother's unit fed us in their mess tent, and put us up overnight on spare cots in one of the Army tents. A strange feature of station facilities was the stools in the latrine. A number of two-foot diameter brown glazed clay sewer pipe tees were joined end to end in a row. One-foot diameter, very short nozzles faced up on the line of tees; the sides of the nozzles were flush with the surface of the sewer pipe. The very narrow "seat," provided by the horizontal rim of the nozzle precluded sitting for spell. Water running along the bottom of the tees continually flushed the "stools." We spent the night visiting Foiles' brother and his buddies. The next morning, after chow, we were on our way back to Howard Hall—thankful that our living conditions were much better than those of his brother's Army Unit.

Sometime later, Foiles' brother got a pass and came up to Yarnfield to visit for a while. Foiles and I were both on duty at the guardhouse and Howard Hall gate late that night, so his brother spent the time there with us. It was an amazing thing because Foiles' brother was a bard; much of his conversation was in verse. He could rhyme just about anything we talked about.

MILTON COOK

One afternoon, a GI arrived at the Howard Hall Gate to visit his sister, who was one of the WACs stationed there. Telephone calls were made to locate her and give her the news that her brother was here. It was poignant, running down the walk to greet her brother with a tight embrace, she emanated such joyfulness that I could feel it. I experienced her delight. It was a profound experience. It was as if there was some kind of telepathic transference of emotion.

CHAPTER 36

Singular Service-Meeting With a P-38

It was just one more of the many overcast, but only moderately cold winter days. Cook was in his room maintaining the shine on his shoes in preparation for this nights patrol duty. The door opened and Miller, who had the dayroom duty, entered the room, "Cookie, the Desk Sergeant called and said that you are to get dressed for duty, take early chow and be at the guardhouse by eleven hundred three zero hours for special duty."

"Okay, do you know what I'm going to be doing this time?"

"Nope, he just said what I told you, nothing more."

Figuring that he was going to stuck out side in the cold some place he decided a field jacket would work better under his overcoat. But, it would not be so cold during the day that he should put on a pair of fatigue trousers under the wool ones. Cook then put on his MP gear, grabbed his mess kit and made a quick round trip to the mess hall. Arriving at the guardhouse, Cook opened the door, walked in and asked, "Okay, Sergeant. I'm here, what am I going to be doing this time?"

"Hi, Cook, I'm glad you got here so soon. Anyway, there has been a plane crash and you have airplane guard detail[41] this afternoon. So, go on out and get in the jeep with Wellington and he will drive you to the site."

They went out and got in the jeep and were off dashing through the countryside and hamlets to where the plane had crash-landed.

---

[41] The 1257th was responsible for guard detail, until recovery was completed, for any U.S. Army Air Force Aircraft crashes within a designated area in the English midlands.

Wellington turned the jeep onto a side road, soon stopping at a gatehouse where a security guard logged the jeep in to the facility, and then he pointed to the airplane sitting in plane sight on the ground off to the side of the of the road from the gate to the factory, saying "There's your aircraft." The factory was located in a large other wise empty slightly sloping deep grass field. About a quarter of a mile from each of its sides, a concertina barbed wire topped cyclone fence established the facilities boundary. The wide-open fields were an indication that probably either dangerous or secret work was being performed in the factory. The expanse of factory windows had been "blacked out" by coating their inner surfaces with apple green paint. About half way up the road to the factory and about a hundred yards off from the road, there it sat—a P-38 Lightening fighter[42]. The aircraft had made a wheels—up crash landing, and other than bent props, it did not appear to be damaged very much. The airplane recovery team would have to remove the outer sections of the wing in order to truck along the country roads.

"Okay, Cookie. I guess that's your baby over there," Wellington exclaimed as he stopped the jeep to let Cook off. "Someone should be here to relieve you at about sixteen hundred hours. Have fun, you lucky guy."

Cook got out of the jeep, "Thanks for the ride, Wellington. At least it's daylight and not too damned cold." The jeep went on up the road to turn around in the parking lot as Cook started hiking over the slightly bumpy, ankle deep grass to the downed aircraft. He walked around the plane, looking at everything and thinking, "*This is really great, getting to see a P-38 this close up.*" With the fuselage sitting on the ground, he could climb over the top of one of the twin booms where it was narrowest just in front of a vertical tail; the lower part of which had plowed into the ground and was probably damaged. Now that he was between the booms, he could look through the canopy into the cockpit and just see the upper part of the instrument panel, which he could not have done if the aircraft had been sitting on its landing gear. He could have climbed up on the wing to look inside the cockpit, but he did not think that he should. The aromatic odor of aviation gasoline permeated the air around the plane, so it was

---

[42] See endnote 12 for P-38 Lightening fighter parameters.

probable that running out of fuel was not the reason that the P-38 had crash-landed.

After his initial inspection, there was nothing to do for the next three and half-hours, except to walk around or lean on some part of the downed aircraft. During the whole time on duty guarding the P-38, he was alone and never saw anyone enter or leave the factory. Finely, at last a jeep pulled up to and stopped at the factory gate, and Cook was on his way, reaching the road just in time to intercept the jeep. As Benton, his relief got out of the jeep, Cook said, "Boy, am I glad to see you guys. Welcome to nowhere."

"Any special orders for this post, Cookie?"

"No, Benton just the usual, keep the bystanders away from the airplane. It's a good thing that you brought your flashlight because soon it's going to be very dark as well as very boring. Enjoy yourself, have fun with you very own airplane." Cook got into the jeep with Burris, "Let's go!"

"OK, Cookie we are on our way. I think that you had the good shift." And he drove on up to the parking lot to turn around, and then they were on their way back to Howard Hall.

As the jeep sped along the country roads, Cook thought, *"I'm really glad that I was only out there with that P-38 during the day when I could look it over really good and did not have to spend four hours by myself in the cold and complete darkness of the night."*

Chapter 37

Battle of Britain Parades

On August 1, 1940, Hitler had issued the order "the German Air Force is to overcome the British Air Force with all means at its disposal, and as soon as possible." D-Day, *Operation Sea Lion*, the invasion of Great Britain was set for September the fifteenth. In the ensuing air battle, the Luftwaffe suffered very heavy losses, and with heavy losses, the RAF still survived. On September the seventeenth, Hitler postponed the invasion indefinitely, and the Battle of Britain was over. The RAF was victorious. Thereafter, September the seventeenth was celebrated at Battle of Britain Day to commemorate the victory.

Now, in retrospect it may appear to be strange to be celebrating a victory over the enemy while the war still raged. But, when Germany failed to destroy the RAF and so had to call off the invasion of England the chance was gone; because at this point in the war the tide of battle was at ebb. So this victory was not only a significant victory for the British but was also a victory for civilization. Because after all I have written this and I have written this in English.

While the 1257th Military Police Company was in England on September 17, 1943, having just arrived in late August, it was not really established as a local entity, and so was not invited to be an element in any celebration. By September 1944, the men of the Unit were a familiar sight, a part of the local scene, and well known in the local communities where we were on patrol every night. So it was not surprising that the Company would be invited to march in one of the local parades to be staged in celebration of Battle of Britain Day. The 1257th would add some color and provide a display of American collaboration in the war. It was also fitting as there had been, in the RAF, Americans[43] who were flying, fighting, dying, a few of the few, during the Battle of Britain. Perhaps, this too had been the finest hour for these airmen. Our Unit was, of course, a sharp and showy enough outfit that we could march with the best of the British Units, which would also be in the parade. Oddly enough, this coming parade would also be the first time we had marched extensively as a Unit since August of 1943 when we completed our training at Camp Ripley.

BATTLE of BRITIAN DAY PARADE, ECCESHALL

As Eccleshall was a small community, the parade would be short, made up of a small number of Units and without a lot of loud, interspaced and out-of-sync bands. The staging up moved ahead relatively smoothly, but there was the usual confusion that is always a part of a one-time event like this march. It was a short promenade along the main street to the conclusion of the parade at St. Michael's, a Norman-style church with its squared off belfry. This was Sunday and therefore the parade ended at the church for the Battle of

---

[43] See endnote 13.

MILTON COOK

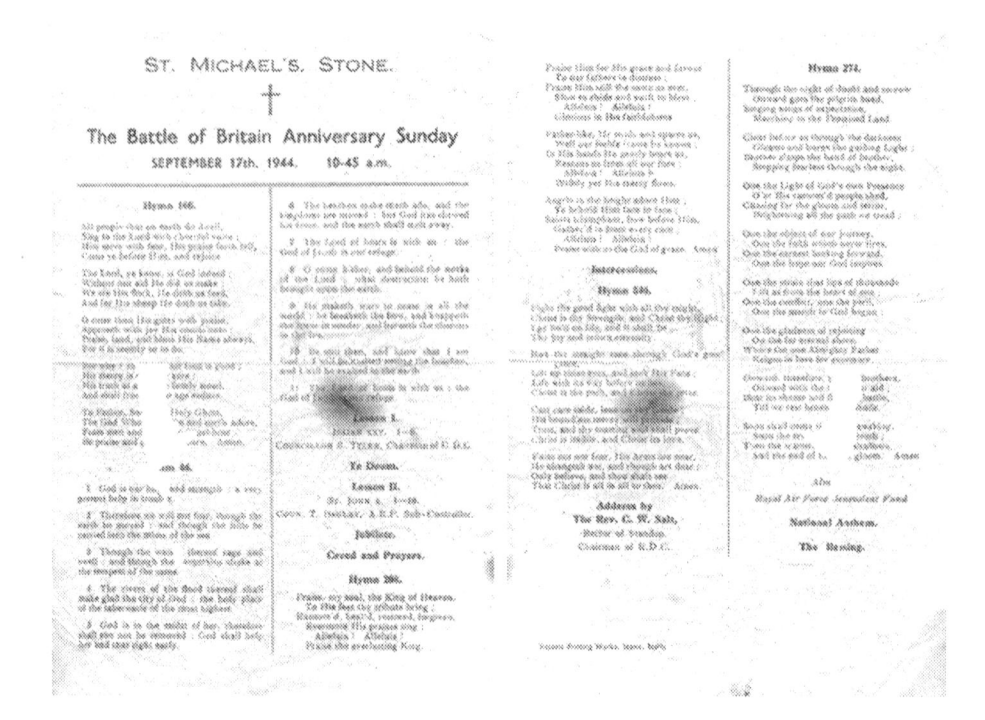

PRGRAMME for BATTLE of BRITAIN SERVICE

Note printers error, most likely did not have enough space for "Eccleshall" with the font used.

Britain anniversary service. Still a part of the parade, we filed into the church with our loaded side arms, removed our white helmet liners and, in a military manner, filled several rows of benches on the left side of the center aisle. I noticed that each of the carved stone capitals were different on each of the cylindrical stone columns in the church. The carvings on the back and sides of the wooden benches held our interest until at last the service started. It was a typical Anglican service except for the address by the Reverend Salt. After completing his remarks pertinent to this day's celebration, he took advantage of this special opportunity to single out the American soldiers for special attention. With the mindfulness, doubtlessly typical of that expressed from pulpits time and time again through out the ages, he chastised us, the foreign soldiers, for blemishing English womanhood in general and the young ladies of the village in particular. While at the same time he overlooked the fact that they were willful participants in whatever activities he had in mind. This was particularly ironic, and we felt that we had been unjustly insulted as, when on patrol night after night, we had to deal with the same subject, so we were in part on his side.

After the blessing, we got up and filed out of the church, formed up in ranks and marched back to our waiting trucks. We were back at Howard Hall in time for lunch. A few men had the rest of the day free, but for most of the company, it was gate or patrol duty sometime later during that day or during that night.

One year later, the Battle of Britain celebration in Stafford was a very different occurrence. The march on this day would be very much longer, both time and distance, and would have many more Units than the parade the year before in Eccleshall. It was early in the morning, the temperature typically cold, and the sky was overcast, when we dismounted from our trucks. The Company was assembled in ranks and then marched over to its assigned position on a very large playing field, which was covered with deep, wet grass. As we stood at ease, in position, we became colder and colder with each passing minute. Time passed slowly as we watched the confusion of the other Units arriving and being located in their positions on the field. Some of the arriving Units appeared to us to be kind of strange and were a source of interest that helped to pass the time. I was particularly fascinated by the trousers, which the young boys of one unit were wearing, boys who probably were enrolled in some local military grammar school. Instead of the usual vertical creases, up and down the front and back, their pant legs had horizontal creases alternately in and out, which were spaced apart by a length that was about the width of the trouser leg. I speculated that the boys must fold the trousers so that they would be compact enough to store in a small locker, but odder still, the creases were so sharp that they appeared to be the result of pressing. It was likely that these creases were' as it is so often the case, an artifice of a process that had become an enhanced traditional form. Too late, even before we had joined the parade and were at last were on our way, we had become aware of the Universal Law of Parades. Which is (and no one would ever have told us) that you should never ever have anything to drink in the morning on the day in which you are going to be in a long parade—particularly if it is on a cold day. Finally, we were on our urgent way marching past the thronging crowds lining both sides of Stafford's main street.

MILTON COOK

BATTLE of BRITIAN DAY PARADE, STAFFORD

When, after what seemed to be a very, very long time, the parade at last came to an end and was dispersed. The 1257th separated from the disintegration and marched on until it entered a large area that was paved with smooth concrete. The Company came to a halt about fifty yards from a facility where lunch was to be served. At the order "Company Dismissed," a charge was mounted, and we were off on a headlong dash for the latrine. This visit was by far the highlight of the day. Afterwards, when we were in the chow line, one of the ladies, who was serving us the food remarked, "My, you boys must be really hungry the way you came running when you were dismissed."

"No ma'am," I replied, "It wasn't that we were very hungry, it was an urgent need to dispose of some liquid that led us to make a run for it!" This was only our second and was to be our last real parade, because the Company was evaporating, and we all would be on our ways to home by the end of the year. So this important lesion about being in parades, which we had learned under pressure, would be down the drain!

CHAPTER 38

Tales from the Barrack-Crossing Paths

On the morning of June 6, 1944, soon after we were out of our bunks, even before morning chow, we heard and were excited by the news broadcast of the Armed Forces Radio. It was D-Day-Operation Overlord, the invasion of the continent was underway. For all of us, in a way, this was fabulous news because now our almost—so it seamed—endless waiting was over. Now at last, we knew that sometime, yes, sometime in the future, we would finally go home. The current details of the invasion we would find out about later in the day, when the Stars and Stripes Army newspaper arrived at the station.

It had been two days past two years ago that I had graduated from high school, and now quickly following D-Day, I was to see two of my class-mates. It was just a few days after the invasion of France when I had an unexpected visitor. A best friend and high school buddy, Bob Vernon showed up at my room. He knew from my letters that I was stationed here so he had made inquiries to find out the location of my room. His flight crew had just ferried a new B-24 bomber across the Atlantic Ocean, via Greenland and Iceland, and had landed at the Burtonwood Air Force Base only the day before. The crew had turned the bomber over to an Eighth Air Force Quartermaster who would then have it flown to a supply depot. After being separated from their bomber the just arrived aircrew, were flown south to Seighford Airfield, which was close by Stafford. After being fed the airmen were bussed to the Repple Depple for paper work and a short time billet. The major part of the air-crews gear would travel by train and be transshipped through Stone Baggage to their assigned air base. The crew would be on their way the next day or so, heading southeast as a new or replacement crew in some unit of the Eighth

Air Force, where they would be assigned a used B-24 to fly and to drop things[44] in the support of the invasion. Bob and I had only a few hours together as I had to go on duty and his crew would be on their way the next day. Vernon was at this point, the only person that I had crossed paths with, and knew from before my induction in to the army. I had parted company with Vernon in March of '43, when he was leaving for the Army Air Corps radioman school and I was off to Miami Beach for MP training. But now he was one of the waist gunners in the bomber crew. Bob completed his missions, returned to the states, and was stationed at Norton Air Force Base in San Bernardino, California (so near to home) and then was discharged from the Army before I ever returned to the U.S.

It was several days after my short visit with Vernon, when I was walking along the passageway between the casual officer mess hall and the station movie-theater, on my way to go on duty, that I encountered Chuck Britten. How strange, after the fifteen months that I had been in the Army, to have this chance encounter with another of my high school classmates in such a short span of time. Britten had been the student body president during our senior year, and the last time I had seen him was on the night when we had graduated from high school. He was now a B-24 bomber pilot, who with his crew had also flown a new airplane to England, arriving just the day before. Then, probably some time tomorrow, they would be on their way as a new or a replacement B-24 bomber crew at some Eighth Air Force Airbase, flying a used airplane. Britten completed his missions, returned to the United States and was separated from the Army Air Force before I returned to the U.S. It had indeed been a strange turn of fate to see two friends so close together at this time and then no one else ever after in the ETO.

These two high school class-mates of mine had, for about fifteen months, been in training to obtain all of the technical skills required for aerial combat. While, I on the other hand had spent only six months in training the skills required of a military policeman.

---

[44] Usually bombs, but on some later occasions, food was dropped to civilians.

Chapter 39

A Singular Comrade

One of my buddies, a platoon mate, whom I would sometimes be partnered with to walk patrol or stand guard at one of the gates, was technically an enemy alien. An uncle, who lived in Wyoming, I think, had sponsored him so that he was able to emigrate from Germany. He was then still a citizen of the Third Reich when he received his greetings from the President of the United States. Upon his induction into the Army of the United States, he took an oath to support and defend the constitution of the United States of America. This was not withstanding the fact that at this point, he was now in the Army, and thereby subject to the articles of war, which placed restrictions on some of the constitutional rights. Bill Amrhein was one of the soldiers assigned to the 1228th and later 1257th when they were first formed in Miami Beach.

At night, once we were in England, there was a lot of time to kill, whether we were on patrol or on gate duty. Both of us were about the same age, initially in our late teens, and about the same height and stature. In addition to the usual discussions about the events of the war, there were many other things to converse about. For one thing, we had very different backgrounds—I had lived in a desert, and he was from a green country of farms and forests. Now, here we both were in England. But by far the most unusual was Amrhein's experience as a member of the Hitler Youth before he was able to emigrate from Germany.

Sometime in 1944, the United States Congress enacted a law that allowed all of the aliens who were serving in the armed forces and who had served for a specified minimum time to be awarded American citizenship. This, of course, was very good news to Amrhein because now he would not have to wait until he had completed the naturalization process to become a U.S. citizen. Arrangements were made, and soon he was headed south to Birmingham, a German citizen in the U.S. Army. At the American consulate, he said, they brought out the book of citizenship papers and used the very first Certificate to change him into a citizen of the United States. It was a good day for Amrhein as he headed back north to the Repple Depple as a brand new American.

Sometime after May 7, 1945, VE Day, Amrhein, based on his particular language ability, was "recruited" to be a member of the U.S. Army of Occupation in Germany. As I recall, he signed on for a regular three-year hitch in the Army with part of the deal to receive a promotion, and so he departed the 1257th. So, now he would be returning to his original homeland, but as a soldier of the victorious U.S. Army. In a liaison function Amrhein would be able to visit family and see friends from the time before his departure from Germany.

What a strange hand the fate of war had dealt to him. Instead of being unknown enemies on opposite sides, we had become friends on the same side. So, is that all there is to it—a quirk of fate, which determines just who will be friends and who will be enemies?

BILL AMRHEIN

The autograph is some—what reminiscent of a high school year-book. This maybe some thing that comes from German culture.

Chapter 40

Singular Service-Tour of the Town

The invasion of Europe was well ahead of schedule, and by August 31, 1944, Patton's Third Army had advanced into eastern France. The Army had passed Rhemes and was approaching Verdon and Nancy when it stalled. The Third Army had literally ran-out-of-gas, its fuel tanks were empty. The Third Army had advanced so fast and so far that it had outrun its supply train. The logistical imperative for a mobile mechanized army is a continual supply of fuel, food and munitions, with the most critical being fuel. As the invasion was ahead of schedule, there were not enough long-range trucks landed in France, and those available could not come up fast enough to supply all of the gasoline needed to continue the advance.

A command decision was made to airlift fuel and supplies to the Third Army to get it back on its way to Germany. There was some information in the Stars and Stripes that the Third Army had slowed because of supply problems. It was when the regular workday was over and those soldiers who had been issued a pass had left the station, that I was instructed to get on an U.S. Army bus and have the driver take me to Stone. It was a school bus of the time except instead of cadmium yellow it was painted olive drab, with white-stenciled U.S. Army markings. Once in Stone, I was to check all of the pubs to find any air transport flight crewmembers and to have them return with me to the Repple Depple. Apparently, some air transport crews had just arrived in England and had been turned loose on passes before the order had come down to send them on as fast as possible as the need was critical. While conducting this grand tour of the pubs of Stone, was somewhat exciting and a little bit of fun, it was not successful because I could not find even one transport crewmember.

It may be that the mission to Eccleshall did better at rounding up flight crews than I did on my excursion to Stone.

Even before any aircrews could have arrived in France from England, the resupply of the Third Army was underway. Located at the time on a forward fighter air-strip in the eastern part of France, Alvin "Bud" Anderson,[45] (not to be confused with triple ace Clarence "Bud" Anderson) pilot of a Northrop P-61A Black Widow[46] Night Fighter, recounted that he had watched in amazement as the resupply activity continued all day long. Douglas C-47 Skytrain transports were landing and GIs were unloading 50-gallon drums of gasoline, rations, ammunition and other supplies into trucks backed up to the cargo door. He said, that just as soon as a truck was loaded, it drove off of the air-strip and headed on down the road and was quickly replaced by an empty one. And, when all of the cargo had been off-loaded, the C-47 took off and soon another one was sitting on the runway with an empty truck by its side.

When the sun had set, the air supply procession had to call it quits. Soon, the airstrip was completely dark, so now it was time for the night fighters to take off into the wild black yonder. Bud Anderson and his crew and the rest of the flight crews who were hanging out at the airstrip climbed up into their Black Widows, fired them up and took off into the night. It was time to prowl so they were off hunting for German bombers, transport, tanks, trains all of which might be moving about in the black of night, and for any other German facilities and forces that they might discover, shoot up and maybe bomb. These Ninth Air Force P-61 Squadrons were real "fly by night" outfits!

---

[45] Our paths had briefly and, of course, unknowingly crossed when he had passed through the Repple Depple (which he referred to as Stone) on his way to one of the Ninth Air Force Night Fighter Squadrons. Bud Anderson related this eyewitness account of these events as a fellow member of American Legion Post 133, Huntington Beach, CA.

[46] See endnote 14 for parameters of The Black Widow P-61A Night Fighter.

Chapter 41

Entertainment-The Theater

The central structure of Howard Hall included the station theater, the "EM" lounge, administrative offices and various mess facilities. The two wings containing the side-by-side "EM" and Casual Officers mess hall and the Howard Hall Theater, covered about the same amount of ground as they stretched to the west. The hallway, which started at the entrance to the Casual Officers mess hall, ran past the row of offices and then passed the "EM" lounge just before it ended at the station movie theater. As the hallway came to its end, it turned right into the entrance to the theater. The motion picture projection booth was located on a second floor just above the end of the hallway. Auditorium seating consisted of very hard wooden folding chairs that were set up in one large central and two small side sections. The chairs could be folded up and stacked along the walls to create an open hall for dances or other like activities. Stage curtains concealed the retractable motion picture screen. The orchestra pit was protected by a squat wall, which was topped with a very short guardrail. The stage and pit were used when an occasional USO show stopped by, but otherwise was not much used. Bob Hope and his USO troupe had come and gone, never to return, before the 1257th had even arrived in England.

The theater was a primary source of entertainment as motion pictures were screened almost every night. The price of admission was low. As I recall, it was on the order of about two shillings (40 cents). Most of the movies were fairly recent Hollywood releases, and they were shown for several consecutive nights; which allowed those of us with duty at night a chance to see the show. The first time I ever saw Danny Kaye in "*Up in Arms,*" I was so fascinated with his antics that I had to see the show every night on which I was off duty while the movie was still being shown. Occasionally, maybe to supplement the movies were then available, an old black and white "classic" would show up on the projection screen. I recall seeing *None but the Lonely Heart* starring Ethel Barrymore and Cary Grant and *It Happened One Night*[47] starring Claudette Colbert and Clark Gable.

HOWARD HALL THEATER

> I never knew until I looked at this photograph that the device painted above the stage was in fact the number 70. An appropriate number as this stood for the some times 70th Replacement Depot and then for the 70th Reinforcement Depot.

The "EM" lounge and snack bar was on the office side of the hallway, just before the entrance to the theater. The lounge occupied a wing, off of the hall that ran east towards the south end of the MP barrack. The floor of the "EM" lounge was raised one foot higher than the floor of the snack bar. The step up was at the edge of the

---

[47] 1934 Academy Awards sweep: Best Picture, Best Actor, Best Actress, Best Director, Frank Capra and Best Screen Play, Adaptation, Robert Riskin. The movie is set in authentic contemporary 1934.

area that skirted passed the serving counters and kitchen area. Among other things in the lounge, there was a register book in which the soldiers, who were on their way to a duty station, could log their passing through the Repple Depple. We were told to turn to a certain page in the book to see the signature of Clark Gable, who had entered his hometown as Hollywood, California. This was the "EM" lounge and was not to be used by officers. So more than likely when he had passed through the Repple Depple, on his way to his assignment in the Eight Air Force, someone had prevailed upon Gable to meet the workers in the snack shop, and workers had in turn prevailed on Gable to sign the register.[48] Gable's entry was always a source on interest until one day, some damned fool ripped the page with his autograph out of the book.

There was one member of the 1257th whose only assignment ever was to police the "EM" snack bar and lounge. This GI just met the minimum height requirement for induction into the United States Army. He was older, close to forty, so with his height and age, his Draft Board was getting near the bottom of the barrel when they sucked him into the Army. He was by far the shortest man in the Company, much shorter than the next shortest in the 1257th. Maybe someone with a perverse sense of humor, or maybe someone just desperate to fill a quota on that day, had assigned this man to the Military Police, and there he stuck. Being as short as he was, it would have been difficult to perform most of the regular MP duties. So a place had been made for him by creating a permanent duty assignment at the "EM" snack bar and lounge. Trouble could never be expected, as the strongest drinks available were cokes and coffee. Nonetheless, the lounge and snack bar was his duty station. In a strange sort of a way, this assignment utilized the entertainment value of this short, stocky, somewhat older GI with a Queens' accent in an appropriate duty station. Picture a just slightly oversized Billy Barty, decked out as an MP, maintaining order of a bunch of soldiers who are just sitting around relaxing.

---

[48] See endnote 15 for a short essay on Gable's service in the Army Air Force.

AN "ENLISTED MENS" SNACK BAR

Note the bombing mission tally painted on the back of the repatriating airman's authentic bomber jacket.

CHAPTER 42

Singular Service-Orders to London Town

In May of 1997, my wife, Roberta, and I flew to England on vacation. After a seven-day bus tour of the southern part of the island, we were heading north to visit the area where I had spent two years in a strangely different time. Before we began our trip to England, a room had been pre-booked in the North Stafford Hotel (ca. 1873), which is located right across the street from the railroad station in Stoke-on-Trent.

It was Sunday, and there were many travelers waiting to board the train going to Northern England and onto Scotland from New Euston Station. With our luggage, we lagged behind the rush and by the time we were able to climb onboard the train, most of the seats were occupied, except for those reserved for passengers who would be boarding at some station up the line. So, my wife and I ended up sitting in different parts of the car.

The railroads in Britain have been electrified, and the train quickly gained speed, whipping by red brick retaining walls, warehouses and factories, and under and over a multitude of bridges. Soon the train was out of the suburbs and clipping along through the green countryside, heading northwest. Somewhat over half of the way from London to Coventry, the train entered a long tunnel. It had been almost fifty years since I last had passed through this tunnel, and if I had been conversing with my wife, I might have missed it. But there it was permeating the air, the acrid odor of coal smoke. It is likely that most of the passengers on the train paid little notice to this weak, mildly bothersome background scent and did not know where it came from in any case. Having recently read in the *Scientific American* about the phenomenon, I knew that the speeding train was pushing a mass of air ahead of it through the tunnel. This fast moving roiling air was scouring a little of the more than one hundred years' accumulation of coal smoke residue from the tunnel walls. Today's trip was nothing like the first time I had gone through the tunnel, I recalled, with displeasure the nose-biting, eye-watering, throat-clutching coal smoke seeping into the coach as it belched from the rumbling steam engine.

00000

The door to the room opened and Campbell, who had day room duty, said, "Hay, Cookie, the Desk Sergeant called and said to tell you to come down to the guardhouse."

"Do you have any idea what he wants to see me about?"

"Nah, he just said it was a special assignment and to send you down so he could give you your instructions."

"Okay, I'm on my way." Cook put on his shoes, which he had been shining, got up from the bunk where he had been sitting, put on his field jacket and garrison cap and closed the door behind him as he headed down the hall. It was a cold, gray Saturday morning, and he thought, "I hope I'm not going to get stuck on some guard detail at some remote spot out in the country." Walking quickly, he rounded the end of the mess halls and started down the sidewalk past the theater and the WAC barrack. Nearing the guardhouse, he saw that Goble and Malone were on duty at the Howard Hall Gate.

"What' cha doing down here this time of day, Cookie?" Malone asked.

"I don't have any idea. I was told to report to the Desk Sergeant for some assignment." Cook opened the door and went into the guardhouse. Closing the door behind him, he said, "You wanted to see me Sergeant?"

"Yeah, Cook, you're going on a trip today. You're on your way down to London this afternoon to escort back here some GI that the MPs there have locked up down there. The London guardhouse cells are full, and the Provost Marshal wants to get rid of this soldier as soon as possible. So, that's why you are going down there today. Here are your orders and travel papers, which I just got. Your train leaves Stone at sixteen hundred and fifty two hours. So, you be back here at the guardhouse no later than sixteen hundred hours, ready to go, and someone will drive you to the station."

Taking the set of documents Cook glanced at each one and then started to read the first. Dated today's date, 6 Jan 1945, Special Orders No. 6; *E X T R A C T* ; *Section I*; Line 9., placed him on temporary duty not to exceed two days to go to London for the purpose of escorting a Pvt. James R. Malerich, held in confinement by military authorities, London to confinement at AAF Station 594, which would be the cell in this guardhouse.[49] A paper clip was used to hold together the multiple copies of the orders, which had been typed, on lightweight, eight by thirteen inch sized paper, using black carbon paper. Each sheet of the orders had been crisply embossed with the H.Q. 70th Replacement Depot seal. The train schedule to London had been prepared using the station "standard" mimeographed form. The schedule showed one train change in Stafford, with about a half-hour layover, and arrival in London at twenty hundred and fifty hours. There were two war department railway warrants, one to be exchanged for one ticket from Stone to London, and the other for two tickets from London to Stone. Last, and the most important was a sheet of paper on which someone had been written down the address (and the nearest Underground station) of the specific Army Unit (the London guardhouse) where Pvt. Malerich was held in confinement by military authorities, i.e., military police.

"These orders say not to exceed two days temporary duty. Am I supposed to come back here tomorrow or on Monday? With this train schedule, I think that it will be at least twenty two hundred hours before I get to wherever it is that I'm going."

"You can stay down in London tomorrow and bring your prisoner back here on Monday. I expect that they would have a problem on Sunday processing the paperwork and providing your transportation to the railroad station. Monday morning should be soon enough to satisfy the London Provost Marshal."

"Okay, I'll be back here by sixteen hundred hours for a ride to Stone," Cook said holding the papers in his hand as he opened the door and started back to the barrack.

Back in his room, he thought, *"I might as well get packed for this trip,"* getting his musette bag out of the locker. Packing for this trip he placed in the bag, his toilet kit, two changes of underwear and

---

[49] See endnote 16 for essay on a promotion mystery.

socks, a towel, a armed forces edition book (that he was currently reading), a English Mars bar and a roll of Lifesavers (both from his this week's ration), one of his MP bassards, his Sam Brown belt with side arm, and the set of documents (except for the railroad warrant for his ticket to London).

It was time to head out so Cook, put on his service jacket and slipped the railroad warrant into the inside pocket. Next on was his overcoat and service cap, and then picking up his bag, he was off on his way to the guardhouse. Walking in the door, he said, "Well, here I am, Sergeant. Ready to go. Who's going to take me over to Stone?"

"Oh yeah, Cook, You've got your orders. Good. Bunker, take the jeep and haul Cook down to Stone station and then get right back here. No screwing around."

"Let's go Cookie." Bunker said as he headed out the door. "So, where are you off to this time of day, Cookie?"

"I'm on my way to London to bring back some GI who they have locked up down there."

"Hey, that sounds like a good deal. How long do you get to stay down there?"

"Just one day, Sunday. My orders say two days maximum. But, I'll get there so late tonight that we figure that means forty-eight hours."

Bunker stopped the jeep at the bottom of the hill, where the road ends at Highway A34 to talk to Kovacs, who was on traffic duty at the intersection. "Hey, where are you guys headed this time of day?"

"I'm hauling Cookie to the train station. He's on his way to London this afternoon."

"Oh yeah? Why are you going to London, Cookie?"

"They are sending me down there to escort back some GI that they have locked up. AWOL, I guess."

"What a deal. How long will you be there?"

"Just one day. I'll be back here some time early Monday."

"Well, have fun. Talk to you guys later."

The jeep spun around in empty parking lot in front of the Stone railroad station and braked hard to a stop. "Thanks for the ride, Bunker."

"No problem. You have fun in London, Cookie." Bunker said as he let the clutch out of the jeep.

Stopping at the outside ticket window Cook presented the warrant to the agent, who after taking some time to read it, finally handed back a ticket to London. Once he had the ticket, there was nothing else to do so he sat down on one of the benches to wait. It would be about twenty minutes before the train to Stafford would arrive. As he waited, on his way all—alone once again, he thought, "*Well, here I go off to a strange place, and this time it will be dark when I get there. I hope that I can get to where I have to go without getting lost. I wish I wasn't going down to London so late in the day. Hell, at this time, I just as soon not to be going down there at all.*"

The train came grumbling and hissing to a stop. Watching for a passenger exiting from a compartment, he quickly found an empty seat. He put his bag behind his feet as he settled down for the quick trip to Stafford.

The train tracks and the platforms in the Stafford railroad station where completely covered over with a roof of glass panes. These glass panes had all been painted over for the blackout, so there was always some lighting in the station. Cook had been in this station many times for a cup of tea when on patrol duty. But, it would only be about a half an hour before the train to London would arrive and so there was not really enough time to get something to eat in the station restaurant.

The coaches on the London-bound train were the long-trip kind with a side passageway running the length of the car. Windowed doors closed off a series of compartments, with facing upholstered benches like those in local coaches, but shorter by the width of the passageway. It was winter, so the clocks were only set an hour ahead, but even so it was dark, and all of the shades had been pulled down to black out the windows. Not being able to view the scenery, Cook had removed the book from his bag before stowing it under his seat. This train was an express and would make only a few stops in the larger cities, such as Birmingham and Coventry, as it rushed south to London's Euston station. Soon after entering the railroad tunnel, south of Coventry, he had to stop reading for a while as he eyes were burning from the coal smoke.

The train slowed and squealed to a stop at the end of the line in Euston station. The passengers stood up, gathered all of their belongings, crowded into the passageway and gushed from both ends of the coaches—engulfing the platform in a flood of humanity. Most of the travelers appeared to be in a hurry to be on their way home or to wherever. But Cook took his time, as he did not know just where he was going or what he would do when he finally got there. Trailing behind the crowd, by this time almost alone, he made his way to the Underground station. Selecting one of the vending machines, he determined the fare to the Bond Street station. He dropped the required coins in the slot and out popped a ticket. At an adjacent machine, there was a very drunk, very young British Brigadier[50] fumbling away as he tried to buy a ticket from the machine. This was all very unusual because while full colonels were fairly common, it was a very rare occurrence to chance upon a General-grade officer, especially one so soused. This was a welcome to London town, which he would never forget.

Arriving at ground level from the Underground, Cook walked out into the almost deserted Bond Street station. Looking at a map of the local area, he thought, *"Let's see what's the best way to get to Five North Audley Street? I'll go down Davies Street, turn right at Brookstreet, and it will be just past Grosvenor Square. It's the third street that I turn on. Okay, I should be able to find the place in the dark."* It was really dark out on the street. After a few minutes, his eyes adjusted and he could see well enough to start walking down the deserted street. Multi-story buildings lined both sides of the street, homes or flats, with all of their windows dark. Not a light to be seen, it appeared as if there was no one at home anywhere. This was Mayfair, which had been one of London's more fashionable pre-war residential areas. If it had been daylight, he would have seen the fronts of the buildings, which were for the most part faced with gray stone, and all of the all of the wrought iron fences and rails and ornate doorways with large doors festooned with gleaming brass fittings.

Concealed in some of the buildings off and around Grosvenor Square were all of the elements of the United States Army. Enlisted men's barracks, officers quarters—with accouterments and space increasing with rank, NCO and officers' clubs, enlisted and officers' mess rooms, dispensaries, mailrooms, post exchanges, many offices, a guardhouse. In short, any and every function required to operate a modern industrial army[51] was to be found somewhere in these buildings.

---

[50] Equivalent to an American Brigadier General
[51] That was a mostly a non-profit enterprise, which was staffed for the most part with indentured labor.

Shutting the door, he turned and went over to the desk. "Hello, Sergeant. I'm Cook, they sent me down from the Seventh Replacement Depot in Stone to escort back there some GI who you have locked up here. These are my orders."

The Desk Sergeant reached over and took the Orders from Cook's hand. "I'm sure that someone knew that you would be coming here today, but no one passed on any info to me about you coming here tonight. Just let me check and see if anyone was aware that you would show up here tonight." Turning around to face the Duty Officer, who was sitting at a desk behind him, "Lieutenant Wimberley, Sir, do you know anything about PFC Cook's Orders and that he would be showing up here tonight?"

"Let me check, Sergeant. There was a stack of Teletype messages on the desk when I came on duty. I looked through them, and there was nothing in them that required immediate attention. Okay, here it is," and then he proceeded to read the entire set of Orders out loud.

"Well, we have received a Teletype of your Orders, but I'll just keep a copy to log you in from," said the Desk Sergeant, while handing back to Cook, the remaining copies of the orders. "This is a hell of a time to show up here. Have you had any chow since noon?"

"Only a candy bar and some Lifesavers that I brought with me. There wasn't enough time when I was between trains, and I did not want to try to find some place where I could get something to eat—if I could—after I arrived here in London."

"Well, that's the Army for you. Those birdbrains, who make up the Orders, have no consideration for the guys who have to carry them out. We've got a small mess in the other room for the men who will be coming in from patrol duty any time now. Why don't you go in there and get yourself something to eat. When you're through, Corporal McBride will get you situated with a bunk and show you where the latrine is. There is no way you could get anyone out of here tomorrow morning, so enjoy your day in London. But, you better be—back here in the guardhouse Monday morning by eight hundred hours, ready to go. But who knows how long it will take before you are on the way back north."

"Thanks for the invite. We also have a little chow when we finish up patrol duty at night. I'll be back in here, ready to hit the sack just as soon as I have some of your chow."

It was about six hundred hours, Cook was awake and thought *"Might as well get up. What the hell is that?"* The rumble and the roar of a long freight train arcing through the air across London terminated by a thundering explosion as the two thousand-pound warhead on the V-2 rocket[52] detonated. Welcome to the new age of ballistic missiles. *"Well, nothing to do now but get up, clean up, and find our where I can get some chow."*

Before starting off to see the sights of London, he put his Orders into the inside pocket of his service jacket. In the unlikely event that he was stopped and checked by an MP, the Orders would establish that he was, in fact, on duty and therefore did not require a special pass to be in London. It was a crisp, clear day. A marvelous day to be sightseeing, to visit some of those places in Mayfair and Westminster that he knew about from the radio, books and the movies. Before it became dark, he returned to the temporary billet for supper. After eating, he went out with a few of the local GIs for a little entertainment.

Monday morning, back from breakfast, it was time to get ready for this day's duty. He adjusted the MP brassard so that it fit snugly around the sleeve before putting on his coat. After putting on and adjusting the Sam Brown belt and sidearm, he picked up his bag and headed for the guardroom. It was just seven hundred hours when he stopped in front of the Desk Sargent. "Hello, Sargent. I'm Cook, I'm here to escort Private Malerich, who you have here in custody, to the Seventh Replacement Depot this morning".

"It will be a little while yet, Cook. We have a couple more GIs who are also on their way out of here this morning. After all of the paperwork is completed, we are going to haul all of you to your various train stations in one truckload. You won't be here long so grab a seat, and I'll let you know when everything is ready for you to go."

Parking his butt on a straight back chair against the wall, Cook was soon joined by two other MPs. All three MPs would soon be on their way, shepherding three sad GIs back to where they belonged.

"Okay, Cook. I'm ready for you." He got up from the chair and walked around to the front of the desk. "Now, Cook, Malerich's

---

[52] See endnote 17 for parameters of the V-2 Rocket, and additional comments.

personal property is in this envelope, so sign here for receipt and then sign here on this form for his custody. Okay, you are all set now. Here are your copies. Malerich, you are now in PFC Cook's custody, so you will do just what he tells you to do. Both of you can go out now and get up in to the back of the truck. You'll be on your way as soon as I transfer custody of the other two GIs."

Cook placed the envelope and the copies of the receipts in his bag and put the strap over his right shoulder so that the bag would hang on his left side. "All right Malerich lets go on out side to where the truck is parked."

"Okay, Malerich, you climb up in the back of the truck, and I'll be right behind you." After climbing up himself, Cook sat down on the bench on the right side of his prisoner. When the other two MPs and their men had boarded the truck, the driver showed up and closed the tailgate.

"All aboard! First stop, Euston Station. Enjoy your trip, wherever you're headed." Another truck driver who thought that he was a comedian. The traffic in London was mostly buses, taxis, delivery trucks (lorries) and military vehicles. There were a number of stops, starts and turns as the truck made its way through the streets of London. After one stop, the driver showed up at the back of the truck and dropped the tailgate, "Whoever is getting out here, we are at Euston Station."

Swinging down to the pavement from the back of the truck, Cook said, "Okay, Malerich. You can get out of the truck now." As Malerich was exiting from the truck, Cook turned to the driver, "Thanks for the ride. I'm glad we survived."

"Don't mention it," the driver replied swinging up and latching the tailgate. Then he climbed in the cab of the truck and rushed off leaving them standing in the street in front of the station.

"Okay, let's go, Malerich. First, I've got to locate the ticket counter and get our tickets back to Stone," Cook said as they passed through the monumental stone block framed entrance[53] to the railroad

---

[53] Though now no longer dirty from years of coal-smoke the side frames of the original entrance now stand in front of the new Euston Station; two monolithic columns in a small patch of grass. All that remains of the original station.

station. After exchanging the second railroad warrant for two tickets, Cook selected a bench to sit on until their train was announced. Once on the train, they found a compartment where they could set side by side. The trip back north was more interesting than the trip down to London as he could now watch the countryside slide by and see the railroad yards and stations in the cities. And this time, he knew just when the train entered the tunnel and the smoke was just as bad as on the trip down.

They had about a half an hour layover in Stafford before the local train to Stone arrived at the station. Watching the passengers exiting the train, Cook quickly located a compartment with two empty seats. Arriving in Stone, Malerich followed Cook out of the compartment and onto the station platform. The other passengers who had just gotten of f the train quickly vacated the platform. "Okay, now we are going to go up and over the tracks on the pedestrian bridge and then take the sidewalk over there down to Stone baggage, so that I can arrange for a ride out to Howard Hall." When they arrived at the gate, they turned and went into a large circus tent where the baggage was stored coming and going. There was only one soldier in the tent sitting at the table reading a copy of *Yank*. "Hello, I'm Cook, we just arrived from London. Would you call the guardhouse and have someone come here and pick us up?"

"Okay, will do. Things are kind of slow here today. That's why I'm here all by myself with no truck."

About twenty minutes later, a jeep pulled into the baggage tent. "Hi, Richards. I'm glad that you could get here so fast. You get in the back of the jeep Malerich, and I'll get in beside you."

The jeep backed out of the tent and turned to head down the hill. "Where have you been off to, Cookie? I didn't even know that you were gone."

"They sent me down to London on Saturday afternoon to bring this guy back here. Not much notice. They just told me I was going, and I was gone. But, I did get to spend Sunday sight-seeing in London."

"Well, that's a good deal. Anyway, some guys have all the luck."

The jeep dashed through the Howard Hall gate passing Denny and Pintar, made a sharp right turn and braked hard along the side of the guardhouse. "Hey, thanks for the ride, Richards," Cook said as he pushed the right front seat forward, turning as he backed out of the jeep. "Okay, Malerich. End of the line! Let's go into the guardhouse and see about fixing you up with a place to stay."

"Hello, Sergeant. I'm back, and I'm turning over the custody of Private Malerich to you now. This is the envelope with his personal effects and here is his transfer form." As on his last escort duty, this was the last he would see of Private Malerich, who would in the course of time be subjected to a punishment that the Army deemed fitting for his breach of good conduct.

"If you don't need me for anything else, I'll take off as they will be closing the chow line soon."

"Okay, Cook. You are off duty for the rest of the day. You are down for the Duncan Hall Main Gate at eight hundred hours tomorrow. Did you have any expenses?"

"Only the Underground fare from Euston station to the Bond Street station that is near to the Army headquarters in London, and that was not enough to bother about."

And with that, Cook headed out of the guardhouse on his way back to his room. He dropped his bag on the bunk, took off his sidearm and overcoat and dropped them on the bunk, grabbed his mess kit from the locker and was off to the mess hall for a late lunch. "I wonder what movie is showing at the theater tonight? I think I might just go and see it."

London had been a fine and some—what exciting experience, it is not every day that you hear a V-2 drop in, but it was finer still to be back here with his buddies.

Chapter 43

Tales from the Guardhouse-Suicide Watch

"Hey, Cookie, time to get up. It's twenty-three hundred hours." Moore who had the day room duty, was waking me up from a less than sound sleep. It was almost time for me to go on duty at the Howard Hall Gate. Whenever I had midnight to four hundred hours guard duty, I would always try to hit my bunk early, but unlike a lot of people, I usually had a hard time going to sleep. But now it was time to get up, get dressed and along with Goffinett, my partner for this shift, head down to the guardhouse. It was a little cold outside at this time of night, so we put our field jackets on over our duty uniforms. We would remove the field jackets as soon as we were in the warm guardhouse, where we would take turns watching the gate through the pass window. Also, walking along with us were Pearson and Lawrence on their way to relieve the guard detail at the Duncan Hall Gate. They were wearing their overcoats, as the gatehouse there would be very cold. There was an electric hot plate on the floor, which was little help against the English Midlands midnight chill. The relieving desk sergeant was also part of this bleary-eyed bunch of soldiers on their way to a late night duty station. As usual, when getting up out of my bunk in the middle of the night, I was in a kind of stupor, feeling very dopey. At least when I got to the guardhouse, I could have a cup of coffee, which might wake me up a little bit. After a while on the midnight shift I usually would start to feel almost normal. But as it was to turn out this night's duty would not be exactly normal.

There had been at least one suicide at Beaty Hall that I knew of during the time that we were stationed at the Repple Depple. I always wondered that with all of the adversity that we had to contend with and still survive up to this point—why now? Why now, well maybe it was fear of being maimed or maybe the fear of a slow and painful death. Also, now being in a war zone the abetting government issued side arms, with live ammunition, were available. At any rate, on this night there was a soldier who by his actions had been perceived to be suicidal, for whatever reason. Because of this GI's predilection to self-destruction, it was decided that an around-the-clock watch had to be maintained as a preventive measure. And what better than to have the MPs do the watching, as some of them would be up all night long in any case. Having the MPs do the watching meant that no one else, medical or otherwise, would lose any sleep over him, and would not have to get up out of a nice warm bed in the middle of the might to keep him company.

It was a custom in the 1257th to always relieve the man who was on duty just a little bit early and never late. But even before I was officially on duty, I was instructed to remove my sidearm so that I could be locked up in the cell to keep an eye, so to speak, on the potential suicide. This would be the only time while I was on duty as a MP outside of the United States, in a war zone, that I was not armed with a loaded weapon. It was also the only time that I was ever locked up in a cell.

So, I removed my pistol belt, with the sidearm still in its holster, and handed the whole set to the desk sergeant for safekeeping. This was very serious business, surrendering the weapon for which you were responsible to the custody of some other person. Now that I had disarmed myself, the desk sergeant unlocked and opened the door to the guardhouse cell. Out came poor Fox who had just spent four hours in the cell, who would very soon be sound asleep in his warm bunk. Now it was my turn, so into the cell I went. The door swung closed behind me, and I heard it lock with a loud click. Now that I was in the cell, I looked around in the dim light and located a very hard wooden folding chair. This seat was close to the bunk where our suicidal soldier slumbered. Sitting "alone" in the dark, with no one to talk with and not even being able to read, time dragged on slower and slower into the extending night. The chair became harder and harder, the ache in my butt increased correspondingly. Still not completely recovered from getting up in the middle of the night, the increasing discomfort was not sufficient to keep me from periodically dozing off. It was just dumb luck that I did not fall so fast asleep that I fell off the chair. Throughout all of this battle to stay awake, the hours of adversity, our pending suicide slept like a baby. He must have overcome his notion of self-destruction, because for the entire time that I sat by him, he never once stirred, he just laid there stretched out on the bunk, dead to the world.

Chapter 44

Singular Service-Time Out With a P-47

Cook was diligently working his way down the hallway, mopping the floor as he went, when Greene abruptly popped out of the day room at the end of the hall. "Hey, Cookie, the desk sergeant called and said to you to take early chow and then report to him by eleven hundred one five hours. You are to be dressed for duty. It's a special assignment."

"I wonder what the heck it is this time?"

"I don't know, he didn't say. I'm just relaying the message to you."

"Okay, well I'll go eat and then get ready as soon as I finish mopping this hallway."

After dumping the bucket in to the deep sink and then put it and the mop back in to the broom closet, Cook took a few steps to his room and grabbed the mess kit from the hook in the locker. Still in his fatigues, he was not allowed into the mess hall until after explaining that, he had been instructed to eat early chow before going out on some special guard duty.

He knew that a special assignment on such a short notice always would be some kind of guard duty. This was a cold winter's day, and he was certain that whatever it was that had to be guarded, it would be out in the open some place. So, it was very important to dress for the weather, but at least, it would be during the day instead of at night. First, replace the fatigue shirt with a wool OD shirt, then take off his shoes and pull a pair of wool pants over the fatigue ones. Next, he put on a second pair of socks over the pair that he had on, and then pulling tops of the folded socks over the bottoms of the trouser legs. *"This extra pair of socks problem won't help very much to keep my feet warm,"* he thought putting on his shoes and leggings. Standing up, he pulled on a wool sweater vest, over which went a field jacket. He draped the gray RAF scarf around his neck before putting on the wool overcoat, with the MP brassard already on the left sleeve. This was the best he could dress to contend with the damp penetrating English winter cold. Fastening the sidearm around the waist, putting the helmet liner on the head and finally picking up a pair of wool gloves, he was ready for inaction. Now bundled up, it was time to make the walk down to the guardhouse. Entering, Cook reported, "Hello, Sergeant, I'm ready for whatever it is that I'm going to be doing this afternoon."

"Okay, Cook. Good. We have a crash-landed aircraft, which we will be guarding until it's retrieved sometime tomorrow. You go on out with Sergeant Celentano, and he will drive you to your post."

"Okay, Sergeant. I'm on my way."

They headed north towards The Potteries, and after a few miles drive through town and countryside, they pulled up and stopped on the outskirts of one of the many communities, which comprised this hub of pottery production. There, in a large wide open, very flat field, only a few hundred yards from the edge of the town, sat a P-47 Thunderbolt. The fighter was resting on its belly, having made a wheels-up landing, and other than sporting the requisite bent prop, it appeared to be in good shape. It was just eleven hundred five zero hours when he walked over to the MP whom he was relieving. "Hi, Grimes. As you can see, I'm here to take your place. Are there any special instructions?"

"Boy, am I ever glad to see you, Cookie. The only special instruction is to keep all of the sightseers away from the plane. So, goodbye! See you later. Have fun." With that, Grimes walked over and swung into the jeep and off they drove, leaving Cook inspecting the grounded P-47.[54]

The onlookers—boys, girls, young and older women and a few men—maintained a discreet distance as they slowly skirted around the dormant Thunderbolt. There was a steady flux of sightseers drifting out from and back to locations scattered among the ranks of two-story, red brick, row houses; all of which ended on a line to establish an even edge to the community. This downed airplane was

---

[54] See endnote 18 for parameters of the P-47 fighter.

a subject of great interest. Probably, most of the spectators had never had a chance to see any fighter airplane up close. So, they were taking advantage of this more than likely sole opportunity to view the P-47 up close, as any of us would. In fact, this was the only time that Cook was up close to a P-47. Anything unusual and out of the ordinary like this airplane laying on the ground will always generate some excitement, but the winter's chill shortened the attention span for the casual observers. For most of the visitors, it was one quick circuit around the grounded bird, and then a brisk walk back to wherever they came from. There were, however, a few hangers-on, who stayed and just looked at the P-47, real aircraft buffs who just could not tear themselves away. Among these last stragglers remaining at the crash site there was a young woman, a somewhat strange appearing young woman with very dark make-up.

The pale winter's sun was sinking, and the shadows of the row houses extended across the field towards the conclave of people and airplane. It was near the end of Cook's shift, and he was hoping that he would be relieved just a little bit earlier than usual. At last the MP jeep arrived and stopped a short way from the crash site and the bitter-end enthusiasts. Cook's relief, whom he saw was The Bad Penny, dismounted from the jeep.

"Hi, Cookie. You can take off now. Do we have any special instructions?"

"Only one. Just keep everyone well away from the airplane. Did you bring a flashlight? It will be night soon, and you will probably be out here alone in the dark, so a light may come in handy. Anyway, have fun." Cook turned, walked over and climbed into the jeep along side of Hubbard, who immediately drove it away.

The next day, Cook learned that The Bad Penny had not returned to Howard Hall when relieved at twenty four hundred hours. Instead, he had been picked up when the guard was changed at zero eight hundred hours the next morning, having spent the intervening portion of the night with the strange young woman. It was very apparent that he had acted on my counsel to "have fun."

Chapter 45

Entertainment-The P.A. System

Radios were a scarce article in the Repple Depple. The 1257th had one located at the guardhouse. It was an American radio to which an adapter for the higher British voltage had been attached. Also, the power connector had been changed to a standard British one, which was inserted into a recessed socket and held in place by screwing down a captive-retaining ring. After twenty hundred hours when almost all station activity had ceased, those who were on duty at the guardhouse would listen to the radio to help pass the time. On rare occasions, we would be able to pick up the broadcast from some station in the U.S.A. But usually, we would be listening to one of the U.S. Armed Service radio stations broadcasting in England. Some of the programming of the Armed Forces radio station originated locally, but most of what the station broadcast were programs and music recorded on V-disks. The V-disks had been cut strictly for the use of the U.S. Armed Forces serving outside of the continental United States. At this time most of the music on V-disks was never released and sold as a commercial record. Also, music continued to be recorded on V-disks when the musicians were on strike against the recording companies and were not making records for the American public. These recordings on V-disk were frequently not the very same version as those, which were sold to the public in the States.

In the afternoon, once all of the official work of processing personal through the Repple Depple was completed, and the mess halls were open for supper meal, music would be played over the PA system. The music played included all of the songs then popular back in the United States. The windows in my room faced the back of the mess halls where one of the PA speakers was located, so I could hear whatever was being played over the PA system at the time loud and clear. When I was in my room, usually after chow, getting ready for some night duty, the music was there. The exception was when a heavy rain was falling or a strong wind was blowing, the music was not played over the PA system.

Whoever was spinning the platters, must have had his very own favorite records, or maybe there were a lot of requests for some of the numbers. At any rate, every day for long periods of time certain records would issue at least once from the speaker. I remember, in particular, day after day hearing Spike Jones and the City Slickers' unique versions of *Cocktails for Two* and *The William Tell Overture*—which degenerates in to a horse race. Conversely, there were V-disks that were little heard and probably should not have been played at all. For instance, there was a live recording of the then young Frank Sinatra with the audience of screaming adolescent girls, which was not popular at all. Nothing against Sinatra, it was the particular audience, which did not sit well. This particular V-disk did generate some adverse comments.

A special category of popular ballads was the song of fidelity, sexual fidelity that is. Throughout the whole of World War II, there were recorded many of this distinctive kind of anthem. In all of these songs of fidelity, the onus was primarily on the women to be true. It is not at all clear to me just what the social imperative was that made these songs popular. Maybe some reinforcement of the sentiment of fidelity was necessary to maintain some measure of cultural steadfastness. These tunes were more than likely all written by men, but were widely accepted by women. The fact that men were usually not implicated in these ballads may be because there was a well-founded uncertainty as to their fidelity (this was after all in the time of the double standard). Certainly there was little incentive or intention for many to be faithful when bountiful opportunity to be otherwise presented itself repeatedly. One time, I heard a young English woman exclaim on this subject, "The married ones are the worst!"

Early in the war, in 1942, while I was still in high school, we listened to the declaration of fidelity, *No Love, No Nothing (Until my baby comes home)*. Later on we would take heed of *Don't Sit Under The Apple Tree (With anyone else but me, 'til I come marching home)*, which was a warning that one had better be faithful. Then there was the declaration "I'm true to you by default" as expressed in the song *They're Either Too*

Young Or Too Old (I don't set under the apple tree). Late in the war, and I think the last song of fidelity, the poignant *I'll Walk Alone*, which declared "I have no problem being faithful, but it's really the pits being here without you."

There were a number of recordings of *I'll Walk Alone* featuring various vocalists, of which three were particularly noteworthy. In the United States, the musicians had been on strike again against the recording companies. RCA had yet to sign with the union, so in the U.S. a hit rendition of *I'll Walk Alone* was the a cappella version of the song recorded by Dinah Shore with a back up chorus. But at the same time there were other versions of the song recorded, by the musicians on contract to the recording companies that had settled with the union, and on V-disks. In the Pacific Theater of Operations, the most popular rendition of this song was by vocalist, Martha Tilton, who vied with Dinah Shore for the hit U.S. recording. Her version, I think, was the stand out interpretation of the ballad. While in the European Theater of Operations, we were caught up by a still different transcription of *I'll Walk Alone*, which issued from the P.A. system for a long, long time. This particular V-disk recording of the song was by Louie Prima with Lilly Ann Carol as the then band vocalist. Lilly Ann Carol was unknown or at least little known by those of us who had been in the ETO for some time. Being relatively unknown and not being one of the established vocalists, we were not listening to just another popular rendition of the song, but were relating to Carol's plaintive declaration of fidelity and loneliness, which exactly fit the mood of that time and of that place.

CHAPTER 46

Guarding the Gate-A Generally Fine Day

It was a bright and clear winter morning early in 1944. On duty at the Duncan Hall entrance gate, Senjanin was inside and I was standing at ease on the concrete pad in front of the gatehouse waiting: waiting to stop and log in any U.S. Army vehicle that entered the Repple Depple; waiting for this duty shift to end; waiting for the war to end; waiting to return to the United States; waiting to be discharged from this Army; waiting to go home. It was very boring, all of this waiting, for very seldom did anything interesting or exciting ever occur.

DUNCAN HALL ENTRANCE (NO. 2) GATE

View from the far side of Yarnfield Road, the gatehouse is on the right behind the tree.

Coming from the direction of Stone, a shiny black Army Staff car abruptly turned off from the road and came through the gateway. It was apparent that the driver did not intend to stop. In fact, the vehicle did not even slow down. It might be said based on his appearance that he had a Hollywood type of insolence, because after all he was a chauffeur to the stars. Just as the automobile turned off of the road past the station, I observed that there were flags streaming from short staffs attached to the top outside edge of each front fender. The American flag, the Stars and Stripes, fluttered from one rod and on the other, there was attached a red flag bearing four large white stars. Perceiving the situation, I quickly decided not to stop and log in this particular car. So instead of my prescribed duty, I snapped to attention and executed a snappy salute. The three general officers sitting in the back returned my salute as the vehicle whipped past me.

Seated in the center was the Major General Carl A. Spatz, Commander of the United States Strategic Air Forces (USSTAF) in the ETO, which included the Eighth and Ninth Air Forces. The Seventh Replacement Depot was part of his direct command as it processed the personnel requirements for the several Air Forces in the ETO. Seated to the left of General Spatz was Lieutenant General James H. Doolittle[55], Commander of the Eighth Air Force, and on his left sat Lieutenant General Ira C. Eaker, currently commander of the Mediterranean Allied Air Forces (MAAF) in the MTO.

This visit of the Generals to the Repple Depple was officially billed as an inspection tour. But I really think that General Eaker was in England at the time, and that all three of them had some free time, so they decided to visit an old comrade. Certainly such an inspection was not in the purview of General Eaker. The Commander of the Seventh Replacement Depot, Colonel Ira A. Rader, who wore command pilot wings, was a fellow member of the small coterie of Air Corps pilots in the Air Force from the First to the Second World

---

[55] Doolittle is a somewhat incongruous name for a man who, aside from his military service, also made significant contributions to the advance of aviation. He was, in fact, Dr. Doolittle, with a doctorate in Aeronautical Engineering from MIT. The March 2003 Flight Centennial Issue of The Smithsonian Air & Space magazine has him first in the profiles in its selection of ten significant contributors to aviation development.

War. General Spatz had been a fighter pilot in WWI and had three "kills" to his credit. So, while it was an "inspection tour," it was also a boondoggle, as there was really nothing in the operation of the Repple Depple, which required the personal attention of these general officers.

It was many years later, in the vicinity of 1980, when I along with some of my fellow workers attended a luncheon, which was part of a machine tool trade show. General Doolittle was a featured speaker in the luncheon program. The General said at the conclusion of his talk that after the luncheon was over, he would be available to whoever wanted to meet him. A fairly long reception line formed, which was moving at a good rate. We were all eager to speak with, and shake the hand of, this hero of aviation, and so we fell in at the end of the line. When it was finely my turn to meet the General, I told him about when our paths had crossed briefly so long ago in England. He said that yes, he had been stationed and served in England during the war. I, of course, knew that he had in fact been minding the store, even if he did not say as much. But, it was clear to me that while I remembered well our quick "passing", he had no recollection of the event. Ha!

COLONEL RADER, GENERAL SPAATZ AND
GENERAL DOOLITLE

CHAPTER 47

Entertainment-A Furlough At Last

It was January 1945, the 1257th had been stationed in England for one and a third years, and now those of us who did not have relatives in the UK were also allowed to have a short furlough. So, taking advantage of this new policy two of my buddies, Foiles and Strehlaw, and I applied for and were granted a seven-day furlough at the same time. In wartime Britain, and particularly with the buildup of military forces, there were very few places to go to which were more or less away from the proximity of anything military. And there were very few such places that also had available lodging and some type of entertainment. Somehow, we selected Blackpool as the place to spend our short respite from MP duties. Located north of Liverpool, on the western coast of England, Blackpool is primarily a summertime beach resort. Being as it was a vacation (AKA Holiday) spot, the town had resources for amusement and many places in which to stay, all of which, I think, led us to the selection of Blackpool as the spot for our get away.

It was on a Sunday morning when the three of us showed up at the guardhouse to get a ride to Stone Station in the MP jeep. Soon after arriving at the guardhouse, we learned that a shutdown had been ordered and so all passes and all vehicular traffic, except for the MPs of course, were suspended. As prescribed for these occasions, the MPs would set up roadblocks to intercept any errant trucks and their drivers, and MP patrols would be sent out to apprehend any stray GIs who were wandering about AWOL. We were assured that the shutdown did not apply to furloughs, so we climbed into the jeep with Jarnagin for a fast ride into town.

SOLDIERS THREE
COOK, FOILES, STREHLAW

Heading north from Stone, the train quickly passed through Stoke-On-Trent, in the Potteries, which were the prescribed normal limit of our range when on a pass. Now, we were really on the way to somewhat unfamiliar and different part of the country. After leaving Manchester, the next major town was Bolton. Here, we had a few hours layover before the train to Blackpool was to depart. So we decided, "What the heck. Let's walk around and have a look at some of this town and kill some time." While crossing an expansive stone-paved city square, we were intercepted by a pair of MPs on the hunt for any AWOL GIs. They checked our furlough papers, which of course established that, we were legally out and about, and so told us we could be on our way. But we soon had to head back to the railroad station to catch the train to Blackpool.

Soon after departing from Bolton, the train passed through Chorley. This was sort of a homecoming as only a little over a year before, we had been for a time, on temporary duty at the Chorley station of the Repple Depple. I had been on that very railroad station platform on so many nights while on patrol duty in the town. Shortly before arriving in Preston, the train passed through Bamber Bridge, which was the location of yet another station of the 70th Replacement Depot. After leaving Preston, the train had only a short journey to reach the end of the line in Blackpool.

Arriving in Blackpool, the first thing to do was find a place to stay for our short period of "freedom." As this time of year was really off-season, it was not a problem to find a bed and breakfast with individual bed and bathrooms. It was cold this time of year, and of course, the rooms were not heated. Though the blankets were heavier than we were used to, there were enough on the bed to keep warm as long as you stayed under them. It was quick to bed at night, and in the morning, there was no desire to lounge around in the bedroom. It was get up, clean up, get dressed and get out as quick as you could. This was not too different from what very many GIs in England had also experienced. What made this experience special was sleeping between the sheets in a bed, which was longer and wider than I was. And that also this was the only time that I was to sleep in a real bed while in the Army.

In the morning, we would have a typical English breakfast, including such things as kippers, bangers, fried eggs, oatmeal porridge and always lots of toast and jam and several pots of tea. After the completion of this for us very different meal, it was nice to sit around for a while enjoying the nice warm room with its coal burning fireplace. We spent the time reading the newspaper and whatever else was available. This was a real vacation.

After a while, bored with all of the leisure, it was time to head out and roam around the town searching for whatever might be interesting. One of us had a camera, so on one of our outings, we took pictures of each other, both two-by-two and individually. Passing by the almost deserted beach, the cold green surf rolling in from the Irish Sea, breaking and slithering up on the sand, did not look at all inviting. This time of year, most of our time would be spent in an arcade, a sizable building bisected by a large, enclosed atrium with bridges cross connecting the surrounding gallery. The arcade was warm and contained all manner of shops—tea and lunchrooms, a cinema, a dance hall—in fact, the whole gamut of places needed to have fun. An unusual attraction was a large glass fronted coin-operated musical automaton. This music machine was an intriguing show to watch with horns, a trap drum and a violin going through their routines whenever they were paid to perform.

The week had passed, and our time was up. So, we boarded the train and headed south, back to Stone and to duty. This had not been a very exciting furlough, nothing like being at home with family and friends. But, it had been a good break from the old Army routine, which at this point was what it had been all about. This was the only furlough which Foiles or I would ever be able to take during our almost three years' service in the U.S. Army.

OFF SEASON HOLIDAY IN BLACKPOOL

Every member of the United States Armed Services was entitled to thirty days furlough or leave, depending on the branch of service, for each year of service. But, the sorry fact was that very many, maybe the majority, of those who were serving their country were not able to utilize all of the "vacation" time to which they were entitled. In fact, the state of affairs was so bad that soon after the war was over, there was much clamor about compensation of the former "enlisted men"[56] for the earned, but unused because it was unattainable, leave. To resolve this problem, the United States Congress reluctantly decided to make restitution for the unused but earned time off from duty. Special Armed Forces Leave Bonds were issued in $25 increments, to provide compensation for the due but not taken furlough and leave time. Not counting the effect of inflation, the bonds, if held to

---

[56] Commissioned officers were paid for their unused leave time. Presumable the rule was that "enlisted men" were required to take all of their earned time off. Use it or loose it.

maturity would provide a slight overpayment for the unused leave. (But, how many would hold the bonds to maturity?) By issuing bonds, the problem was thus solved without the congress actually authorizing the funds, which would be required to pay off the veterans, and so the magnitude of the embarrassment was never disclosed. I have no idea if the heirs received anything for the due but unutilized leave for those who fell along the way.

It is interesting to note that the maximum mustering out pay, regardless of length of service, was $300. I estimate that I was due, as a Private and PFC, a little over $150 for my unused furlough time. I was issued a leave bond with a maturity value of $175. So, it appears that a very large sum of money was involved in this unpaid leave case.

Chapter 48

Singular Service-Guarding the Gate at Nelson Hall

It was a little past six hundred hours and almost everyone in the barrack was out of his bunk. Only the few who had just finished duty at zero four hundred hours were still in their bunks, as they were allowed to sleep in. The latrine was a scramble of activity, men showering, and shaving, getting ready for the day's duty. Mirostan, who had the Day Room duty, slipped his way past the mostly unclothed bodies looking for Cook.

"Oh, there you are! Cookie, the desk sergeant called, and he said that you are to be down at the guardhouse ready for duty at seven hundred three zero hours."

"Okay. I wonder what the heck it's going to be this time. I thought that I was going to be on patrol duty tonight."

After completing his clean up, Cook went to his room and put on his OD uniform instead of fatigues. Now it would not take too long to finish dressing for duty when he returned from the mess hall.

"Hello, Sergeant. What the heck have you got me doing this morning? I thought that I was scheduled for patrol duty tonight."

"Yeah, Cook, you were going to be on patrol duty tonight, but I have a problem that you can help me take care of. They are short a man at the Nelson Hall this morning, so you will be going over there for the morning shift. You will be working with one of the men who are on regular duty there. You can go on out with Birr, and he will drive you over there in the jeep."

Millmeece was only about a mile away, so it was a very quick trip to Nelson Hall. The morning's duty at the only gate was the routine, with only a few vehicles passing through, which had to be logged in or out, and no foot traffic. A little after eleven hundred hours, the MP, who had regular duty on this gate said, "Cookie, the chow line is set up by this time in the morning, but the mess hall will not start serving food for about a half hour. One of us on duty at this gate normally goes over and eats early chow at this time of day. So why don't you head over to the mess hall and eat the early chow today."

"Okay. You can hold down the fort while I'm gone. When I'm back, I'll relieve you and then you can go get your chow. But you will have to tell me just where the mess hall is, and where the door to it is located."

The directions were simple enough, and Cook was soon inside the empty mess hall. Walking along the wall towards the chow line at the far end, it suddenly hit him that he didn't have his mess kit with him. How was he going to be able to eat some chow without a mess kit? A long steam table, filled with many large pans of what was lunch, ran across the building separating the kitchen area from the dining area. Nearing the line of chow, he saw that the wall on his right had an opening and that an identical line of chow faced another eating area on the other side of the wall. The second mess hall was almost identical to the one in which he was now standing. The principal difference was that the other mess hall had some racks filled with stainless steel, three-compartment mess trays, round containers filled with eating utensils and trays of handle less cups. Well, this solved the dilemma of the missing mess kit. So grabbing a mess tray and some utensils, Cook served him self lunch from some of the food-filled pans. Both of the side-by-side mess halls were still empty; he was the first to have lunch. But the one that he was in was the closest, so randomly picking a spot at a table near the front of the mess hall, he sat down to eat lunch.

Eating alone in the large table-filled hall and in any case being armed it was not necessary to remove the white MP helmet liner from his head. Well into lunch, Cook looked up to see a commissioned officer scrutinizing him. He smiled at the officer who in turn kind of smiled back. This officer, who was probably the mess officer, and who of course did not recognize this strange MP, shortly went on out of the mess hall. Finishing his lunch, Cook placed the used tray and utensils in the dirty service rack near the front of the hall. Going back through the opening between the pair of mess halls and then out of the entry door, he walked back to the gatehouse to relieve the other MP so that he could go and eat his lunch.

It was some time later, after returning to Howard Hall, that it dawned on Cook (but did not particularly concern him) that he had eaten in the casual officers mess hall. It was okay, but the chow had not been any better than the food, which was usually served in the "EM" mess hall. This had been a different kind of experience. There were few, if any, of his comrades who could say that they had dined, and dined alone, in an officers' mess.

NELSON HALL
In this view the main gate is just off to right.

NELSON HALL FROM THE AIR

This view of Nelson Hall shows the typical "H" configuration of the Repple Depple barracks. By chance my last view of the Repple Depple was of Nelson Hall, which was still there, in July 1947 when I passed by on the train on the way south to London.

CHAPTER 49

Occurrence in the Barrack-A GI Party

In the army, there was an extreme vexation, the fallout of the great gathering of citizens from diverse regions and cultures and, now and then, some from just plain poor upbringing. It was a state of affairs, which to my knowledge has never been a footnote to the history of World War II, at least until now. In every unit, there had to be at least one man who filled a particular niche, an individual who's particular personal hygiene fell well below the socially acceptable margin. It was almost essential that there had to be at least one member of a company who exemplified the lower limit of malodorousness.

The singular position as the **very most hygienically deficient**, was occupied in the 1257th by "Piss Poor Precision" (this was not exactly his real surname). Now he earned his sobriquet because of his deviant use of "piss poor" in place of the more socially acceptable "f-word" as an all purpose pronoun and adverb. It was probably because it was alliterative that this GI was almost always referred to and addressed using all three of these "p-words."

Piss Poor Precision was billeted in the other wing of the barrack, and this was in a different platoon with different duty assignments, so we in our wing of the barrack were not really aware of his descent into unacceptable nuisance ness. However, those members of the 1257th who had to spend time on duty with him, or even just passing by him in the hall, became fed up with "PPP's" ever-increasing rancid aroma.

The time finally came, I was told, that some of his "closest" comrades decided that a GI Party was in order. That is, action had to be implemented to stress to "PPP" that he must change his ways and "clean up" his act. As the story goes, the volunteer work detail first removed his less than fresh uniform, and then constrained him to lie prone on the floor in one of the shower stalls. In this position, he was painstakingly scrubbed all over using a stiff bristled GI brush and the good old brown GI soap. This was the same soap used in the mess hall to clean the grease from all of the food—encrusted utensils. After the completion of this very thorough scouring, "PPP" was inserted into a "fart sack," the open end of which was then tied closed above the showerhead. After a discreet period of time, he was considered sufficiently rinsed and was released on his own recognizance. It appeared that he had gained some hygienic initiative as a result of this training session, as it was reported that he never returned to his old ways.

Who really knows just how many thousands of times this basic ceremony, with all of its possible variations, was performed during the course of World War II. Generally, this rite had only to be executed but one time to achieve the desired "good" conduct. The GI Party was a very fine example of very, very tough "love".

CHAPTER 50

Guarding the Gate-Cold Feet Again

The winter of '44 was cold, damp and dreary, with fog and frost to break up the monotony. That December, all of Northern Europe was blanketed with thick, low-hanging clouds. These impenetrable clouds kept the fighter aircraft sweating it out, unable to rise from the advanced airstrips and provide air support for the advancing army units. This was the time, December the sixteenth, that the German Army launched their last counteroffensive into the Ardennes Forest in Belgium. This final push by the Germans was to be known as the "Battle of the Bulge."

DUNCAN HALL (NO. 1) EXIT GATE

> Covered with a very unusual dusting of snow, the gate-house is on the left. The two gates were normally closed at night, but would be opened for any late night transient by vehicles.

On this very morning, the fog was lifting as I came out of the barrack and started on my way to Duncan Hall where I was posted for gate guard duty. The countryside, the buildings, the walkways and the roads were all white, rimed over with thick frost. After passing through the Howard Hall Gate, I could see up close that the fence wires had doubled in size. What were normally "bare", one-eighth inch diameter, galvanized iron wires had become quarter-inch diameter white crusted cables. The needle embellished, hoarfrost coating on each wire added to the very odd appearance of the world on this very strange morning. Even with all of the frost, there was still the damp, penetrating cold to contend with.

It was on one evening during this same time of the year when I was on duty at the Duncan Hall Gate that I got cold feet again. It was cold that night, maybe just a little below freezing. There I was standing on the cold, cold concrete checking the passes and ID's of the stream of GIs and officers going through the gate. After a little while, my feet started to get cold. Then, they became colder and colder as time went slower and slower. At first, it was just a dull ache, but with time, the feeling in my feet had progressed to acute pain. Now, there was a hot plate on the floor of the gatehouse, which under normal circumstances, and with the door closed, would provide an illusion of being a space heater. But on this night, there was no help from the hot plate. I tried warming my feet by placing my, leather soled, shoes one at a time on the crimson coils of the hot plate. However, I was so concerned that I might some how burn the surface of the soles of my shoes that I could not keep them on the red-hot coils long enough to do any good. I was very desperate, but not to the point of stupidity. When at last my shift was over, I was able to limp back to my room and eventually restore my feet back to normal. This was another time when I could have put those arctic overshoes to good use.

As it was, this night's session of standing on my own cold two feet clinched a decision that I had already made. Apparently, replacements for combat losses, of MPs at least, were slow in arriving from the United States. The causality rate might have been higher for MPs as they would be posted in more exposed positions in order to direct troop movements. In any case, a notice was sent out asking for volunteers to join up with the army that was now fighting its way across Europe. Now, just for the asking, you could get the fate of war to deal you a brand new hand. While there were some few soldiers who spent the whole war in the good life in the U.S., we, in England, were close enough to the actual action to have some empathy with those who were actually engaged in combat. Maybe it was some sense of guilt, sleeping in a warm bunk in a warm room and eating hot food in a warm mess hall, while not too many miles away, other GIs were sleeping in cold mud and eating cold K-rations.

The very next morning, after this cold night, was the time of final decision. We had to choose between joining up with, or—not joining up with the infantry. I thought, why would I put myself in a position where I could get my feet too cold for comfort on a regular basis. However, the overriding reason that I did not volunteering for this particular duty was quite different. A few days before I was inducted into the Army I had returned to my home in Adelanto to see and tell family and friends goodbye. During the visit with my father, he had told me to never volunteer for anything, for he knew full well the hazard of volunteering. My father had acquired the dubious distinction of being the youngest totally disabled U.S. veteran of World War I. He had enlisted in the U.S. Navy when he was only sixteen years old.

On the other hand, my best buddy, Strehlaw, opted for a re-deal from the fate of war. When he left the Company, I was a little sad. I was going to miss him, but we each must follow our own distant drummer. Before leaving, he gave me his copy of *Target: Germany*[57] (which I still have) as it would not be very practical to keep the book in a barracks bag that would be subjected to who knows what kind of manhandling. Strehlaw was the one comrade with whom I maintained some form of contact over the years. After the end of the war, on one occasion, he said that he thought the reason he had kept his corporal stripes was due to the manner in which he reported upon arrival at his new outfit. During basic training, we had been instructed that upon arriving at a new assignment, you must approach the commanding officer of the unit, salute[58] and state your rank, last

---

[57] This book was sold in the PX. Subtitled *The Army Air Forces' Official Story of the VIII Bomber Command's First Year Over Europe*.

[58] In the U.S. Army, it was proper to salute while uncovered; that is, inside of any type of shelter where head covering was not normally worn, it was still a requirement to salute on such occasions.

name and "reports as ordered." Strehlaw said that there were quite a few military police NCOs, sergeants of various ranks, who probably did not report correctly and, as a result, were reduced in rank to private. Also, it is probable that there were no openings in the T.O. for their specific ranks, and because of the manner of reporting no allowances were made. At any rate, this bunch of soldiers had to be some unhappy S.O.B.s. After volunteering for combat duty, they ended up getting busted, which was not a just reward for answering the call to this more arduous service.

CHAPTER 51

Tales From the Barrack-Troop Movements

While I was never completely aware of how and why it was occurring, there seemed to be a more or less continuous, but small, turnover of personnel in the 1257th. A large core group of the original members of the Company, from the time when it was first formed, always remained together. Yet, some of the original members and even some of the replacement members were replaced by new faces. Sometimes, the necessity of a replacement was apparent, such as the time when one member of the two-man motorcycle patrol had a bad accident. We were never to see this soldier again, but we heard that he had been shipped back to the States. Interestingly, the motorcycle patrol wore British military motorcycle helmets. The helmets were somewhat strange in appearance, looking like a normal motorcycle headgear with a truncated rim British military helmet stuck over its top. While the U.S. Army had big U.S. motorcycles, there were no true U.S. military motorcycle helmets that provided head protection for both fragments and flops.

One of the original members of the 1257th, who had been a detective on a city police force in real life, was transferred to the Army Criminal Investigation Division (CID). He would return to visit us on several occasions when he was in the area on a case of army crime. One time, he stayed at Howard Hall for a short while when he was investigating a death of a GI at Nelson Hall. He told us an account of what had occurred after completion of the investigation. I think that there were other soldiers with special skills, who were also transferred to some other station where they could be of better service to the Army Air Force. At least, there were some other men who had served on a municipal police force that vanished from the company over time.

THE MOTOR POOL

> The motor pool was located in Duncan Hall, but served all of the Stone area Repple Depple units and stations. The entire MP patrol unit is standing by their motorcycles.

For a short time, there was, in addition myself, a man named Baker and a man named Fry answering roll call. I found this muster of three culinary names somewhat amusing. But my impression was that this was not of great interest to anyone else. In any case, at some later time, both Baker and Fry disappeared from the 1257th without my notice.

Another interesting replacement who traversed through the Company was an older man in his late forties, a veteran of World War I. He had been overseas during the Great War because he wore small gold chevrons on the lower right sleeve of his uniform. In the present war a quarter-inch wide by one and a quarter-inch gold bar was displayed on the lower right sleeve of the uniform, for every six months spent beyond the shores of the United States.

One of my comrades, whose name I can no longer place with his face, but with whom I had frequent duty, had been in the Army Signal Corps. He had signed up when a notice had been posted in his unit, soliciting volunteers for military police duty. Replacements were required for those trained MPs who had been transferred to more critical assignments. He said that the MP duty was so much better than the enormous boredom of the endless waiting for the invasion of Europe to get underway. With his new function, there was a lesser level of tedium, and the living conditions were so much better here in the Repple Depple. In fact, the living conditions here were better here than they were in most of the other Air Force Stations in England.

Later on, there would also be replacements for our comrades who had volunteered and departed to join the advancing U.S. Army. I always like to think that Strehlaw's replacement was a GI who had been a member of the Second Rangers, John Saloka. He had been injured while climbing[59] the sheer cliffs at Pointe-du-Hoc on D-Day. Here was a soldier for whom we all had great respect, because he had been there. We, of course, never asked him, and he, of course, never told us anything about what had happened. His injury was serious enough that after he had recovered, he did not rejoin his unit and return to combat. On the other hand, the harm that he had suffered was not serious to the extent that he would have been shipped back to the States. I do not know just what the criteria were for determining whether a soldier who had recovered from battle trauma would return to combat or be transferred to a non-combatant unit. One time, I was talking with a GI who was a tail gunner on a B-17 bomber. He had taken several rounds through the fleshy part of his upper arm. More or less recovered, he was in the Repple Depple processing to be sent back to his unit as an air-crew replacement to finish up his missions. On another occasion, I met a former paratrooper who also told me a permissible story about himself. He had jumped and his main chute had failed to open, so he pulled the ripcord on his reserve parachute. The second chute opened just before he landed in the top of a tree. Straddling a tree limb, his descent was arrested. He passed out and later regained consciousness out of the tree, but was now tangled up in a barbwire fence. So passing out again, he woke up the last time in the in a bed of a field hospital. But now he was in the Army Air Force and so had quit jumping out of airplanes.

While it applied only to our shared military experience in the war, there was an unstated but understood code not to talk about personal incidents. It was understood that whatever assignment you had was

---

[59] Just think about it—another level of excitement to rock climbing, having someone above taking pot shots at you, but with the caveat that you are also armed and can shoot back.

mostly the luck of the draw, and that all jobs were equally important, just that some were more dangerous than others. However, there were some exceptions to the code in that stories were permissible, if the individual telling it was not directly involved in the deed, or if the story concerned something that was silly, stupid or just plain dumb, and was usually humorous. Also, the code did not apply among peers who had shared the same experience. In this case, it was accepted that telling each other war stories about the event was okay.

Only now, as the days dwindle down, has it become permissible to talk about what occurred—to tell some of the stories (some never can be told) in a more or less truthful way, as best they are remembered.

This is a singular story of a transfer that did and did not occur. Sitting on chairs in the guard-house, we were assembled and waiting until it was time to get in to the truck that would carry us to our patrol duty stations on this night. A comrade sitting beside me said that he would be gone soon. He was being transferred to Eisenhower's Headquarters Military Police Company. From what he next said, I figured that he must have objected and tried to avoid the transfer. Anyway, to placate him, I think, they had said that Cook had been the first selection and would have been the one to go except for one item. I had met all of the criteria except for the fact that I was too young. At this time, I was around twenty years old, and the minimum age requirement for this assignment was twenty-five. This was somewhat strange. Here I was considered mature enough to send me off by myself to carry out orders when I was only nineteen. But arbitrarily, I was not mature enough to perform military police duty at SHAEF (Supreme Headquarters Allied Expeditionary Force). I was really very glad that he was the one to be going instead of me. But it is not strange that a person might be subjected to age discrimination at either bound of life.

CHAPTER 52

Tales From the Barrack-Buy U.S. War Bonds

A prevailing English belief was that the American soldiers were overpaid. Probably the major reason for this view was the large difference between the GI's pay and that of the much lower paid British soldiers. A realization developed, particularly after the war was over, that it was not so much the Americans were being overpaid, but rather their British counterparts were being woefully underpaid. In any case, a condition existed where the American soldier did appear to have too much money to throw around. The suggested imposition of a required savings plan for the American forces in Britain, to reduce the amount of their ready cash, was not really feasible. Any such savings program would have to be instituted armed services-wide in all of the theaters of operation. Such a widespread program of forced savings was neither practical nor logical. So, other methods were tried in Britain to reduce the amount of the American soldier's spending in the public eye.

Somewhat similar to forced savings was the encouragement of the GIs to voluntarily have a portion of their pay automatically deposited in the U.S. Army Savings Plan. Additionally, a certain amount of money would be spent on the stations for such things as snack bars, movies, at Commissioned Officers and NCO Clubs, haircuts, tailoring and all of the regularly rationed items available in the PX. Also, the PX sold locally manufactured items, some for use by the GIs, but most were things that could be sent home to the U.S.A. as presents. As the Potteries were close by the Repple Depple, the PX sold fine china products, such as those made by Wedgewood, Royal Dalton and Spode. These items made great presents to send to the folks back home in the States. All of the money spent on the stations would not show up as soldier spending in the towns. At the time, I was not aware of the significance of this fine china, and so I never sent any of it home.

I did, however, send money home to my mother in the form of U.S. Postal money orders. Money orders were really the only method available to convert pounds into dollars in order to send money back to the United States. In fact, the Postal money order was just about the only easy method by which the average soldier could convert any foreign to American currency. Now, there were a few GIs who had managed to accumulate significant amounts of foreign cash, sometimes from gambling, but mostly from black market dealings. The Army soon got wise to the laundering of these illicit funds via the purchase of Postal money orders. So, a limit, based on a soldier's actual pay, was imposed on the value of money orders that could be obtained each month. As I had touched on earlier there was considerable black market activity engaged in by those criminal members of the U.S. Army who had access to the supplies. (A problem that has plagued all armies for all times.) Some of their "profits" must have been spent on high living, which no doubt involved black market "supplies." After the limit was imposed on the value of money orders that could be sent back to the United States, there were more than likely other conduits available to transfer the funds home.

Another method of reducing the amount of spend-able cash in the troops' pockets was to convince them to buy U.S. War Bonds through payroll savings. With my lower pay, and the fact that I was sending money home, these war bonds did not appeal to me. But, apparently, the U.S. Army in England wanted every soldier to buy war bonds. Being a holdout, one of the Company officers spent some time with me, one on one, until I was finally persuaded to authorize the minimum deduction from my pay to buy war bonds. The bonds, when paid for, would be sent directly to my mother for safekeeping until after the war. While I had reluctantly agreed to buy the U.S. War Bonds, I was always aware of the absurdity of the endeavor. Here I was rebating part of my pay back to the U.S. Government, which would pay me interest on the money at some future date. But, this portion of my pay, which I had "invested" in war bonds this month would, so to speak, be returned back to me in my next month's pay to be reinvested in war bonds all over again.

Chapter 53

Singular Service-Orders to Chorley

The Duncan Hall entrance gatehouse telephone was ringing. Walking inside he picked up the handset, "Hello, Cook here."

"Hello, Cook, just the person that I wanted to talk to. After you are off duty, stop in the guardhouse when you come back over here. There is a set of orders for you to pick up. You are being sent out on another trip."

"Oh yeah, where to this time?"

"You're going to be riding with the courier who's leaving tomorrow morning on the regular run up to Chorley."

"Well, that sounds like it might be a little easier and a lot less waiting around than a train trip would be. I'll stop in the guardhouse on my way back, get all of the info and pick up my orders."

When they got to the Howard Hall gate, Cook said to his duty partner, Oliver, "You go on without me. I have to stop in the guardhouse for a little while to find out what I'm going to be doing tomorrow."

Closing the guardhouse door behind him, Cook walked over to the desk sergeant, who had just come on duty about the time that he and Oliver had been relived at the Duncan Hall gate. "Hello, Sergeant, I was instructed to stop in here, pick up some orders, and find out what the special instructions are."

"Okay, Cook, here are your orders. You will be riding with the courier to Chorley tomorrow morning. You and your prisoner will ride back here with him in the same truck on Friday. You are to be over in front of the Duncan Hall Headquarters building at zero eight hundred hours."

"Okay, I'll be there," Cook said and took the orders from the desk sergeant's hand. Reviewing the orders, he saw that the unit's designation had been changed from the 70th Replacement Depot to the 70th Reinforcement Depot.[60] This change in designation was news to him, but it did not change anything that concerned him. The document, which was dated 7 March 1945, was Special Orders No. 66, *E X T R A C T*, item No. 7, placed him on TDY not to exceed two (2) days for the purpose of escorting a Private Bieszozat from confinement at AAF Station 590 to confinement at AAF 594. Both of which were part of the Repple Depple.[61] There were three carbon copies of the order, which were held together with a staple and a paper clip. The orders had been typed on eight-by-ten and three-eighths inch, light tan, translucent paper, using black carbon paper. The three copies of the orders had been embossed together with the "Headquarters AAF Station 594" seal to render them official.

When he returned from the Mess Hall, Cook stopped in the dayroom. "Hi, Scukanne, put me down for a zero five hundred three zero wake-up call. I'll be going up to Chorley tomorrow morning, and I need to be over in front of the Headquarters in Duncan Hall before zero eight hundred hours to catch my ride on the courier truck."

"Okay, Cookie. I've got you down for a zero hundred five three zero wake-up call. Why are you going to Chorley anyway?"

"I have to go up there and escort back some dumb GI, who they have locked up there in the guardhouse, back here to our guardhouse. I'll be back day after tomorrow in the afternoon."

---

[60] The change of the unit name from "70th Replacement Depot" to the "70th Reinforcement Depot" appears to be what is now referred to as political correctness. While the Eighth Air Force had significant combat losses, there were also many airmen who had completed their required missions, who had to be replaced. Reinforcement was not really a better term, as it conveys the idea of weakness that needs to be ameliorated. But it's probable that no one thought of that particular meaning for reinforcement and just wanted to remove the perceived "sting" of replacement.

[61] I continue to use "Repple Depple" even if the expression is not quite on the mark from here on. Maybe, I should use "Reffle Depple," but we never did, so who cares! In any case, Repple Depple was always used, no matter what the designation was changed to on several occasions.

THAT'S THE WAY THE BALL BOUNCES

"Well, that should be interesting, but it sure doesn't sound like a lot of fun to me. I'll see you when you get back."

In the morning, on returning from the Mess Hall, Cook hung up his mess kit in the locker and pulled out the musette bag. This was only going to be an overnight trip, so in addition to his toiletry kit and a towel, only one set of clean underwear and a pair of socks were needed. After placing the side arm, with the belt rolled up and a MP brassard in the bag, he tossed in an Armed Forces edition book. The book was for just in case, not that he expected there would be much time to do any reading. The last thing into the bag was the set of orders. Closing the bag, he picked it up and was out the door of the room on the way to Duncan Hall. There was more than enough time to walk over there and locate the truck, which was going north to Chorley.

THE MAIN HEADQUARTERS

The over all Repple Depple and local station headquarters buildings were located in Duncan Hall. The entrance gate is hidden at the extreme right. The cylinder standing near the center of the photo is a British Post (Mail) drop.

Seeing a soldier standing by a truck parked in front of the Headquarters Building, Cook addressed him, "Hello, are you the driver, and is this the truck that is going to Chorley this morning?"

"Yes, I am, and yes, it is. You must be my passenger Cook. I see you sometimes when I drive through one of the gates, but I don't know the names of most of you MPs. As soon as one more box is loaded into the truck, we'll be on our way."

"Well, I don't remember your name even though I have written it whenever I've logged you through any of the gates."

"Okay, I'm Stisihok, a name which you should remember. But anyway, go ahead and get into the truck."

A short time later, Stisihok climbed into the truck and started the engine. "I'm going to drive around to the Mess Hall first. We'll only be there long enough to pick up the lunches which the cooks have fixed for us."

Stopping the truck at the Duncan Hall exit gate, the driver rolled down his window and handed the dispatch form to the MP, so that he could log the truck out of the station. As the MP returned the dispatch form to the driver, he noticed the passenger, "Hi Cookie. What are you doing in this truck? Are you going AWOL?"

"Hi, Press. Not a chance. I'm on duty. We are on our way up to Chorley. I'm going to be escorting some GI, who is locked up in the guardhouse there, back here to our guardhouse. We'll be back here late tomorrow afternoon."

"You enjoy your trip, have fun Cookie. See you around." The driver rolled up his window, drove through the gate and turned the truck onto Yarnfield Road, heading towards Stone.

Setting high on the right side of the truck in the center of the road, there was a first-class view of the road and the surrounding countryside. Looking out from this high vantage point, he could see over the hedge—rows into the green fields of the farms that lined each side of the road. When the truck stopped at the intersection at the highway A34, Cook rolled down his window. "Hi, Schultz. I see that you are stuck on the point this morning."

"What the heck are you doing in this truck, Cookie? What the heck is going on?"

"We're on our way up to Chorley. I've got to bring back some poor GI who's in the guardhouse up there. I'll be back with him tomorrow afternoon."

"Well, two days just sitting on your butt in a warm dry truck watching the scenery go by is a whole lot better than standing on cold pavement for four hours. See you around. You guys have a good trip, and have fun. The road is clear both ways, so you can take off now, driver."

Cook rolled up his window as the truck made a left turn and headed north on A34.

"Where are you from Cook?"

"California, near Los Angeles, on the Mojave Desert—a place which you've never heard of, not even really a town. How about you? Where's your home?"

"Me, I'm from Columbus, Ohio. A place that I'm sure that you know at least something about."

"Do you make this trip up to Chorley very often?"

"Someone makes a round trip at least once a week. I make the trip up there probably once a month. Most of the time, I'm just hauling, sometimes GIs but mostly just baggage between Stone and one of the Halls. Once in a while, I take you MPs around to the towns where your patrol duty is that night. Occasionally, I get sent to a supply depot to pick up a load of whatever it is that needs to be hauled to one of the Halls."

The truck soon passed through Newcastle-Under-Lyme and then bisected several hamlets along the way. They had passed through Congleton and were approaching Manchester when Cook commented on the fact that there were many people standing along both sides of the road. "The street is lined with people. I don't think that they just came out to see us pass by. I wonder what is going on?"

"Yea, It's very unusual. All of the times that I've made this trip before, I've never seen a crowd like there is here today."

Suddenly a Bobby stepped out into the street up ahead of the truck. He held his right hand up, palm forward, at face level in the universal sign for "stop." "Hello, Yanks. Would you please pull off of the highway right here. You can go down the street on your right and then turn around, come back and park at this corner. I will inform you when it will be alright for you to be on your way."

As the Bobby requested, Stisihok turned the truck off into the side street. After going a short way along the street, he found a good spot where the truck could be easily turned around. Soon parked back at the corner, they had a grandstand seat in a warm truck on this chilly day to wait and to see whatever it was that was going on.

Then, suddenly, there was a black Rolls Royce advancing slowly along the street, the same street that they had been traveling along only a short time before. A livered driver sat in a front space of the limousine, boxed in by a vertical glass separator immediately behind him. In the very much larger rear compartment sat their majesties King George VI and his wife Queen Elizabeth, who were acknowledging the throng as they glided slowly by in the Royal Phantom.

"That was something that I never expected," Cook remarked. "It was kind of exciting seeing the King and Queen of England. It's something to write about in my next letter to my mother."

As soon as the townspeople, who were lining the street just a few minutes before, started to drift away, the Bobby came over to the truck. When Stisihok had his window about half down, the officer said, "Alright, Yanks. You may be on your way now."

"Thanks officer," he said and rolled the window back up before starting the engine. As soon as the side street was clear of pedestrians, the truck pulled out, turning right, and they were on the way again. Passing through the western district of Manchester and then on past Bolton, it was late afternoon when the truck finely entered Chorley. Turning west on the road to Exton, they were soon stopping at the Chorley Army Air Force Station gate—to be logged into Washington Hall.

"You can drop me off at the guardhouse. You will have to pick up me, and the GI who I'm bringing back with us, there in the morning. What time should we be ready to go?"

"Well, I have to pick up the courier box and whatever else Headquarters here has to go south at zero eight hundred hours. It should not take more than ten minutes to load it all up. You should be all ready to go at about the same time. I'll swing by the Mess Hall and pick up our lunches after I pick the two of you up."

"Okay, see you in the morning, Stisihok." "You too, Cook," as he stopped the truck in front of the guardhouse to let Cook get out.

Entering the guardhouse, he walked over to the Desk Sergeant, "Hello, Sergeant, I'm Cook. I figure that you are expecting me."

"Hello, Cook. Yes, we knew that you would be here this afternoon. We don't have any empty bunks in the MP barracks, so we have a spot for you in one of the casual barracks." Turning to the right, "Corporal Ward here will get you some bedding from the guardhouse store and then show you where your bunk for tonight is located. And, Ward, you are responsible for making sure that the bedding is back here in the morning. After you have Cook set up with a bunk, Ward, you can take him to the Mess Hall."

"Good, Sergeant, but I do know where the Mess Hall is. Some of us in the 1257th were stationed here on TD late in forty-three, until you guys got here and relieved us."

"Oh, yeah. I didn't know that. So, you know your way around here then."

"Well, I do know where the Mess Hall is. Will there be any problem getting a mess tray? I don't have a mess kit with me."

"It shouldn't be a problem. Ward will take care of that also. Eat early chow in the morning, and be here in the guardhouse before eight hundred hours. We will have your soldier all ready to go with you at that time."

Pulling two OD blankets and a mattress cover out of the guardhouse—bedding locker, Ward said, "Okay, Cook. Let's go. I'll show you where your bunk for tonight is located."

When they entered the room, Cook was surprised, as it was much larger than he expected it to be. He thought that all of the bunkrooms in the Repple Depple were small. Probably this room had been intended to be used for an office at one time. Anyway, there were six double bunks cramped into this somewhat larger room. A room much to crowded if anyone was to stay in it for more than a few nights.

Ward sat the blankets and mattress cover on the top left bunk nearest the doorway. "You can sleep here tonight, Cook. We'll go to the Mess Hall now. I have to stop by my room and pick up my mess kit on the way."

"Okay," sitting his bag down on the top of the mattress alongside the bedding, Cook turned and followed Ward out of the room.

Returning from the Mess Hall, Cook was surprised to find that the room had filled to capacity with transient GIs. While slipping the mattress into its cover and unfolding the blankets to make up his bunk, he had to provide an account of who he was and just what he was doing in this room at this time. The reiteration resulted in the usual idle raillery about MPs, which was cleverer in the presumption than it was in the reiteration—a part of Army culture.

One of the soldiers in this room was a brand new "GI," a newly enlisted man on his was to a first assignment in the U.S. Army Air Force. Just a few days prior, he had in fact been a member of the

United Kingdom Armed Forces, and now, today, he was a brand new soldier in the United States of America Armed Forces. He, as others before him, had contrived a transfer from the military service of that nation to the military service of this nation, based on the lucky fact that one of his parents was a citizen of the United States. He was discharged from the British Armed Forces and then, immediately, before he could get away, enlisted in the American Armed Forces. Neither nation was going to let him get away with anything. In any case, this Limey was one happy soldier now, with better pay, better chow, better uniforms and maybe even better living quarters. Now, he also had access to the U.S. Army PX, limited of course to his ration allowance. It would be interesting to know just how the British girls viewed those English lads who had transferred into the American Army. Were they now also over-sexed?

All of the other GIs present listened to him with amusement as he exuberantly spilled out a line of BS. He recounted of how he had told those, who shortly before had been his best British buddies, about how good everything was in the American Army. He told them how his Sargent came by every night to make sure that he was tucked into his bunk all right. He was no doubt playing to the notion that some of the British military establishment had that in some ways the American Army was a soft and some—what frivolous organization. At the same time, it was very probable that his past comrades were aware that he was laying it on really thick.

This new GI, being all of his life "English" but now in the U.S. Army, would no doubt always be called "Limey" and would be referred to as "the Limey." This is because there is a general tendency, which is the most pronounced in the armed services, to create nicknames which allude to origins or to surnames based on a somewhat traditional, and sometimes wanting, cleverness. For example, most Native Americans were assigned the honorific "Chief," and much closer to home, "Cook" was usually "Cookie."

Shortly before zero eight hundred hours, Cook, MP bassard on the upper left arm, sidearm slung on the right hip, the musette bag hanging at the left hip and a roll of bedding under his left arm, opened the door and entered the guardhouse. "Hello, Sergeant, I'm Cook. I'm here to take custody of Private Bieszczat and then escort him down to Howard Hall. This roll of bedding is the responsibility of Corporal Ward."

"Yes, Cook. Your man is all ready to go. Would you sign here for his custody and her for his personal effects, which are here in this manila envelope."

After he signing both forms, Cook picked up the envelope with his copies of the forms and tucked them into his bag. Then, looking over at the far wall where the prisoner was sitting in a straight back chair, "Okay, how do you pronounce your name?

"Bi-zat is what they usually call me."

"Well Bi-zat, you can get up off of the chair now. We'll be on our way in just a few minutes."

Almost immediately, Stisihok opened the guardhouse door, "Okay, Cook. I'm ready to leave, so if you are all set you and your man can get into the truck now."

"Okay, Bi-zat. Let's go. I'm right behind you. Go out the door and get up in the truck. You get to ride in the middle." Cook climbed in beside him and put his bag on the floor behind his legs. There was no room on the seat now where he could put his bag for the return trip. After closing the truck door, he said, "Okay, we're ready. Let's go."

The ride back south was not quite as interesting as the journey north had been. They made a rest stop to eat their lunches and later passed through a short stretch of rain. After the truck had turned off of A34 and was on the home stretch, Cook reminded the driver, "You need to go to Howard Hall first and let us off at the guardhouse."

The truck stopped after coming through the Howard Hall gate. The door opened and Cook climbed out, "Hi, Weilandt. You don't have to log this truck in, as soon as we get out of it, he is just going up to where he can turn around and then go over to Duncan Hall."

"Well, Cookie, what the heck are you doing in this truck?"

"We just got back from Chorley. I went up there yesterday to bring back this GI, who will shortly be locked up in the cell."

"Okay, let's go, Bi-zat. You can get out of the truck now. You go ahead of me through that door, and I'll be right behind you."

"Well, here we are back from Chorley, Sergeant. This is Bi-zat. He will be staying with us for a while. His personal effects are in this envelope, and here is his paperwork." As usual, this would be the last Cook would ever see or hear of this poor soldier.

"That's good, Cook. No problem, I guess. I relived of your prisoner and you are off duty now. So, you can go and get yourself some chow see from the duty roster that you have the weekend off. You have patrol duty Monday night."

As he walked up the sidewalk on his way back to the barrack, Cook was thinking, *"Well, that was an interesting trip this time. It would have been kind of diverting to have-gone into Chorley while I was up there to see if there were any changes. Anyway, I'll go on over to Hanley tomorrow afternoon as I have kind of a date with Lillian to do something."*

CHAPTER 54

Guarding the Gate-Action in the STO (Social Theater of Operation)

Mister Roberts, which was written by Thomas Hegger and Joshua Logan, was awarded the Tony in 1947, for the best play of the year. This play about a U.S. Naval cargo ship in World War II is somewhat of a farce. This is not too surprising as, in its own singular way, the U.S. Navy can be somewhat of a farce.[62] While *Mister Roberts* is fundamentally a comedy, it is also in essence a moral drama, which holds the mirror up to life as it bears upon honor—in the sense of respect, duty, fidelity and redemption.

I first attended this play in Norfolk, Virginia around 1950, with John Forsythe playing Mister Roberts. In the middle of the second act, there is a scene with a very obscure bit of shtick. Mister Roberts, Doc and the Petty Officer of the watch, presumably the only three members of the crew who are still onboard the ship, are standing on the quarter-deck at the head of the gangway. Mister Roberts delivers a straight line, "What do you have there, Doc?" Doc reaches into a box, which he is grasping in his other hand and holds up a small packet for all to see. It's a pouch that is so small that only those in the audience who are very close to the stage can see what he is holding in his hand. This olive drab envelope is only about an inch and a quarter wide by two inches long, and maybe a quarter of an inch at its thickest. Doc replies to Mister Roberts, "Oh, just a little present for our boys returning from shore leave." And then, with this line, those who know now know what Doc is holding up in his hand.

---

[62] The author served five years of a four-year enlistment in the United States Navy. Also, see endnote 19 for additional conformation.

Loud, raucous, nervous laughter emanates from individual positions deployed throughout the theater audience. At the same time, most of the audience wonders what is so funny about whatever it is that Doc is holding in his hand.[63] This was, of course a very inside joke, and as more and more of those who were in on it have answered the final roll call, the laughter is fading away. Much as a bit of the topical humor in one of Shakespeare's comedies, this jest has lost its essence and is becoming abstruse. At a recent production of the play that I saw, the lines were faithfully spoken while Doc held up a first aid kit. There were two of us in the audience who chuckled because we knew just what the gag was though unknown to every one else, even the cast of the play.

The packet which Doc was grasping in his hand was just one prong of a trident, prophylaxis, protection and perusal, that Uncle Sam had thrust into service to wage his covert war on VD. "Covert" because back in the U.S. this "war" was not considered newsworthy. This particular weapon was generally referred to as a "Pro-kit" (prophylactic kit).[64] The contents of the P-kit were listed on one side of the pouch as follows:

1. 5 gram tube of ointment,[65]
2. Soap impregnated cloth,
3. Directions sheets,
4. Cleansing tissue.

The P-kit was to be available in all barracks day rooms and at the station pedestrian gates. Also, all MPs were instructed to carry several P-kits with them when on patrol duty. The MPs were to provide a P-kit to any soldier who had neglected to pick one up on his way out and now required a kit for immediate application. Few soldiers, if any, would in the heat of a just completed action have the presence of mind to request a kit from an MP and then use it. In the normal course of events, GIs never asked the MPs for any, and so the transport

---

[63] Now, with the revelation that follows everyone will be in on the joke.
[64] See endnote 20.
[65] A cream made by combining sulfahiazole and calomel.

of the P-kit for emergency dispensing was soon abandoned. There was however a fall-back position in that once back on the station the returning GI could stop by the Dispensary and utilize the always available (no questions asked) prophylaxis unit.

The sheet of instruction in the packet was presumably to refresh the user's memory. But then, who would stop to read these instructions in the aftermath of a just completed action, or, as in the more usual case, read them in the darkness of a blacked-out night. Training in the use of the P-kit was limited usually to only one session at which the instructions from the packet were read aloud and discussed. This was the total and entire extent of the training on the application of the P-kit. This weapon had been deployed for use by the troops without any real hands-on drills or field exercises. With such limited training in the use of the P-kit, there was bound to be many casualties, and there were. A story was related to us about a GI who had contracted the "Big, Big-G." When asked why he had not used a P-kit, he replied that he had. Further questioning established the indeed he had utilized a P-kit, but only after the last time, having spent the better part of a week with his English darling. In the Army, we had a saying, "Some guys just never get the word," which was certainly true on this occasion.

It is most likely that the U.S. Patent Office would rule that the simple removal of everything from the envelope except for the cleansing tissue does not truly constitute invention. But it does appear that just such a revision must have been the precursor of that boon to modern civilization—the moist tissue or towelette.

The second prong of the trident, protection, utilized in the pursuit of Uncle Sam's war on VD was an article of deployed equipment. The condom[66] was utilized in the vanguard of the action rather than, in

---

[66] The U.S. Army Service of Supply (SOS) had to transport from the United States all of the condoms used by the GIs in England. This was because the British condoms proved to be too small. In *Rich Relations*, David Reynolds notes, but does not enlarge on this asymmetry, although Tom Brokaw might have if he had known. A facet of the differing sizes of the condoms, from the two sources, is that it also tends to support the British observation that the American Service men were oversexed.

the case of the P-kit, as a post-action defensive measure. But, there again, the C-device was deployed without sufficient instruction. In this case "none". There were no hands-on training sessions or any field exercises conducted to establish minimum proficiency in the use of the C-device. A recent newspaper article pointed out that even today, in the twenty-first century, some college students do not understand how to properly use a C-device. So what was to be expected during World War II, when Uncle Sam was counting on American GIs' innate know-how and presumed experience to handle this very same "situation." This assumption was made in the face of the fact that at the time, the C-devices were kept hidden away in a drawer under the pharmacy counter. To obtain a C-device, a specific request had to be addressed to the pharmacist or to a clerk using the code name "rubbers."

As with the P-kit, the C-device was stocked in all barracks dayrooms and was available when departing on a pass through the station gates. Also, the MPs were urged to carry the devices when on patrol duty, in the event that some foolish and desperate soldier would request one of them. There never were any requests from roving GIs, and so in time, the MPs stopped carrying C-devices, at least for that particular purpose.

However, some clever member of the 1257th did devise a practical use for the C-devise, which the rest of the company soon emulated. When the device was removed from its olive drab envelope, unrolled and then the ends were knotted together, the result was a very excellent garter, two of which were used to blouse the bottoms of our trouser legs. With this innovation, it was no longer necessary to meticulously fold and tuck the lower ends of the pant legs into the tops of leggings or, later, combat boots. These garters made for very neat and even lower ends on the uniform trousers. The blousing was so uniform and sharp that the bottom of the pant legs sometimes appeared to have been cut off short and then hemmed. The short crisp lower ends of the trousers can be seen in some of the photographs of the men and of the unit, not exactly typical of the military manner. The utilization of the C-device for the fashioning of garters was, of course, the unauthorized use of government property; that is, U.S. Army furnished material was being used for something other than the specifically authorized application.

COLONEL RADER REVIEWS THE 695TH AAF BAND AND THE MPS

In the advanced position trouser legs of the band members are tucked into the tops of their regulation length leggings and are bloused in the prescribed manor. In the rear position the members of the 1257th have bloused their trouser legs using the C-device garters.

MP COMPANY ON PARADE

The 1257th MP Company standing at attention for inspection. The third soldier from the right has his trouser legs tucked into the tops of his leggings. Note the wrinkled state of his trousers. The other members of the company have bloused their trouser legs using C-device garters. Note the sharp creases down the front of their trouser legs. The soldier on the far left has bloused his trouser legs so low that he appears to be wearing spats, which is just a little bit much.

The third prong of the trident, perusal, thrust in to the war on VD was an operation, a maneuver so to speak. One of the salient attributes of a soldier is that they are armed, but whether this has any connection to the title of the operation is not clear. While it was somewhat of a fallacy, in that it might just as well and more rationally termed a "short leg," the United States Armed Forces instituted the short arm inspection.

Therefore, once every month, while the war against dictatorships raged on, the "enlisted" men of each unit were formed up as a group in a line outside of the entrance to the dispensary. Then, one at a time, them would have to enter the dispensary and expose themselves to a medic. The medic would then examine the "arm" for any lesions that might indicate a possible infection of the "Big, Big-S." Next the soldier was required to "strip" the "arm" down, using the thumb and forefinger, to eject any discharge which would in most cases be evidence of a possible infection of the "Big, Big-G."

As American soldiers, we knew that the SA maneuver was a very necessary operation that was required to bring about the defeat and destruction of the oppressive fascist nation of Nazi Germany. And, as free citizens of the Republic, we knew that what we had to do was totally necessary to preserve the dignity of the American way of life. That and the fact that as members of the U.S. Armed Forces, we were subject to military discipline, and as such had no say in the matter of conducting this operation in the war on VD.

Now a question might arise. Were the commissioned officers subjected to the SA maneuver? At least in the case of the Army Air Force officers, the answer was—no. These officers were gentlemen, and as such did not contract VD. Some of them were, however, discretely treated for a social disease. A disease that today would be referred to as a STD (socially transmitted disease).

DUNCAN HALL DISPENSARY

Over all view of the Duncan Hall Dispensary showing the medical staff with little to do on this relatively fine day.

READY FOR INSPECTION

A close up view, on another day, of the entrance to the Duncan Hall Dispensary with a single file of soldiers waiting to enter one by one. What did the original caption, "YOU GUESSED IT" mean?

Chapter 55

Singular Service-Orders to East Anglia Again

The bus pulled away belching diesel exhaust after stopping in Yarnfield where Szucs and Cook got off. They were returning this morning from Hanley where, after seeing their girlfriends off on a bus to Stoke-on-Trent, had stayed the night at the American Red Cross Club. When they came through the Howard Hall Gate, Saxton said, "Hey, Cookie. The Desk Sergeant said to send you in when you returned because he wanted to see you."

"Oh yeah, did he say what for?"

"Nope, he just told me to send you in to see him when you returned."

"Well, I'll see you later Szucs. Looks like I'll be here for a little while." Turning, he opened the door and went into the Guardhouse. "Hello, Sergeant. You want to see me?"

"Yes I do, Cook. You are going to go down to some airfield tomorrow morning to escort a soldier, who is being held in custody, back here. You be here at the guardhouse at zero nine hundred hours tomorrow, ready to leave. Your orders and travel papers will be here for you then."

"Okay, Sergeant. I'll be back here in the morning."

Shortly before zero nine hundred hours, Cook entered the guardhouse. "Good morning, Sergeant. Well, I'm all ready to go. Just where is it that I will be going today?"

"Well, good morning Cook. Here are your orders, warrants and a train schedule. Your train departs in one hour so Walkovich will drive you down to Stone Station in a few minutes."

Taking the documents from the Sergeant's hand, Cook looked each one over carefully. The first, dated 19 Mar 1945, Special Orders Number 78, EXTRACT, Line 33, placed him on temporary duty not to exceed three days to AAF Station 119 for the purpose of escorting Pvt. Edgar C. Church from confinement at that station to confinement at Station 594, this station. The two copies of the orders were held together with a paper clip. The orders had been typed on lightweight gray eight by ten and three eighths inch sized paper with carbon paper, which had seen several uses. Both sheets of the orders had been crisply embossed, with the H.Q. 70th Reinforcement Depot Seal, to render them official. There were two war department railway warrants; one for a ticket from Stone to Stanstead, and the other for two tickets for the return trip to Stone. There was a train schedule listed on the Repple Depple standard mimeographed form. This schedule indicated that there was a sixteen-minute trip north to Stoke-on-Trent. Once there, it would be a fifty-three minute layover before the next train that he would take headed south to London to arrive in Euston Station at fourteen hundred one zero hours. The next train, which he had to catch, would be leaving Liverpool Station just a little over a half-hour later. This looked a little tight as he would have to take the Underground between the two stations, and he had no idea just how long a trip that would be. In any case, he would also have to use the Underground on the return trip when he would have a soldier in his custody. The schedule showed that it was about an hour trip from the Liverpool Station to Stanstead.

"This is interesting. On my other trip to London, I was only on TDY for two days, which I assumed, because I got there in the middle of the night, meant forty-eight hours. And now, on this trip, when I will only be passing quickly through London, they are giving me three days to make the trip. I think that I'll be back here tomorrow."

"Yes, Cook, that's the way the Army does things. You've been in the Army long enough that you should know by now that it doesn't have to make sense. You can go on out with Walkovich, and he'll drive you down to Stone Station in the jeep."

Cook swung out of the jeep, reached in the back and grabbed his bag. "Thanks for the ride, Walkovich. With any luck at all, I'll be back here tomorrow afternoon."

Walking over to the ticket window, which opened onto the platform, he exchanged the railway warrant for a ticket from Stone to Stanstead. Then, crossing over the tracks on the pedestrian bridge, he sat down on a bench. After some time, the train to Stoke-on-Trent finally pulled into the station, coming to a noisy stop. Quickly finding an empty seat in one of the compartments, he sat down, the door was closed by a guard, and the train was on its way again. A quarter of an hour later, Cook exited the compartment and used the pedestrian bridge to cross back over the tracks. He found a bench and sat down for an hour's wait this time until the train to London arrived.

Going along the side passageway, he found an empty seat in one of the compartments and settled in for the three-hour ride to London. About ten minutes after the train pulled out of Stoke-on-Trent, it roared through Stone, where he had started this train trip almost an hour and a half before.

Walking down the platform in Euston Station, he wondered just how complicated it was going to be to get to Liverpool Station on the Underground. When he finally saw the map in the Euston Square Station, he was very glad to find that it was a direct line, without any changes, to Liverpool Station, and only four stops to boot. This was really good news as it would be much easier not having to change lines on the return trip when he would have a soldier in his custody. Once in Liverpool Station, he located his departure gate and waited for a few minutes until the train was announced for boarding.

Finding a completely empty compartment, Cook went in it and sat down. Now there was nothing much else to do, but watch the scenery on this relatively short last segment of this trip. The train had shortly left the city and was rolling along through the countryside when it suddenly stopped and remained where it had stopped for some time. As it turned out, this would be a very intriguing spot to be on a train that was stopped on a siding. Looking out through the windows on the right side of the compartment, he could see a boat canal that ran alongside on the railroad tracks until it turned into a small valley. A red brick bridge with many columns and arches, leaking small streams of water here and there, crossed over the canal and spanned the valley. This bridge was kin to the ancient Roman aqueducts, which it resembled as it conveyed a boat canal from one hill to the next. These two different canals that crossed each other at this point were quite possibly parts of the very same canal. The lower reach of the canal likely curved around after going down the valley a short way and then stair-stepped up through locks to the top of the hill where it could cross over itself.

Through the windows in the other side of the compartment, there was a surprising vista. From the foot of the railroad embankment, where it shouldered into the hill, broad green fields sloped away down to the bank of a wide placid river. Hedgerows parceled out the fields, and spotted here and there were collections of farm buildings. In one of the fields near the river's edge, stood a line of tall, large girth trees. Trees with large knobby tops, festooned with the stubs where each and every branch had been harvested year after year. Up ahead of the train, the river and its valley turned and disappeared behind the high ground on the opposite side of the river. Standing on the level upland in the bend of the river, there was a large factory; an industrial complex engaged in the manufacturing of who knew what.

After sitting at this location for some time, the train started to move again and was soon stopping in Stanstead Montfitchet. Walking out of the station into the parking lot, he saw a parked U.S. Army jeep. A corporal was sitting behind the steering wheel reading an Armed Forces edition of *The Pearl Lagoon*. "Hello, I'm Cook, are you waiting for me?"

"Yep. Hop in and we'll take off as soon as I finish this paragraph."

The jeep pulled out of the parking lot and headed down the road. "Were you waiting for me very long? The train was late getting here. It was sitting on a siding for what seemed to be a long time, to let another train pass by. Sorry to have kept you waiting."

"Hey, no problem. You know that's what we do best in the Army, wait and wait. If I wasn't waiting for you, I'd probably be some other place, waiting for someone or for something else."

"You can drop me off at the guardhouse when we get to the station."

"Yep, that's what they told me to do when they sent me over here to pick you up."

When they stopped at the station gate, he handed his orders to the MP on duty. "Well, Cook, I see that you have come to take away our guest."

Entering the guardhouse, he reported to the Desk Sergeant. "Hello, Sergeant. I'm PFC Cook. Here are my orders which you should already have."

"Yes, Cook, we knew that you were on your way here. We will get you out of here on your way back early tomorrow. So, Cook, you be back here in the guardhouse at zero eight hundred hours, ready to go. We have to billet you with some of the Air Crewmen as they have the only empty bunk in this unit of the Airfield."

"Ramsey, get Cook two blankets and a mattress cover from the guardhouse stores. Then, take Cook over to the barracks where he will be staying tonight."

"These are a bunch of good guys who you are staying with tonight. They will show you where the latrine building is, and take you with them when they go to the mess hall."

The horizontally corrugated arched roof Nissen—Hut looked like a very large tank, which had been cut in half vertically and then tipped over onto a concrete slab. Both ends of this rounded roof building were closed with walls, which had a large window on each side of the central door. This barrack had a row of evenly spaced double bunks on each side of the central aisle, which ran the length of the hut. Shelves, about six feet above the floor, were attached to the sides of the barrack creating a boundary between the "roof" and the "walls." Uniforms on hangers hung from rods attached to the inside and outside edges of the shelves. The floor under the roof behind the uniforms was a storage space for barracks bags and whatever else could not be kept on the shelves. In the exact center of the building, in a wooden frame filled with white sand, there sat a small black iron stove. The black sheet metal stovepipe chimney went straight up and through the top of the hut. These coal and coke fired stoves would do very little to dispel the penetrating cold of the English winters. At night, the light was provided by bare bulbs, which were attached to the ends of wires hanging from the high curved roof.

"Okay, Cook, this top bunk back here in the corner is yours for the night. These guys who are just hanging around in here will take care of you for now."

"Hey, you guys[67], this is Cook, who is bunking here with you tonight. He's an MP who is here to accompany a GI, who's in the guardhouse cell, back to the Replacement Depot."

---

[67] Although not aware of it at the time, Stanstead was an 8th Air Force Strategic Air Depot. These Airmen were not members of combat crews, but instead crewed the bombers that the combat crews had ferried to England from the United States. These Airmen would be members of the crew that picked up these aircraft and flew them to Stanstead Airfield. When replacements for lost or badly damaged bombers were required, they would fly the new aircraft to the requisitioning combat airfield.

"Oh yeah, an MP. You are the first one we've had as an overnight guest. We better be on our best behavior. Where did you come from today, Cook?"

"I came down from the Seventieth Replacement Depot at Stone. I am going to be taking some GI back up there tomorrow morning. I guess that he was a-wall[68] long enough that he will probably spend some time in a stockade."

"That's interesting. So, you are from the Replacement Depot. The home of that den of thieves, Stone Baggage."

"Is that so. I didn't know that about Stone Baggage, but it's not surprising."[69]

"Well, Cook, it's almost time for us to go to the mess hall. Why don't you fix your bunk before we head-out. We'll show you the latrine building on the way there. I expect that you might want to stop there on our way to chow."

After returning from the mess hall, they talked some about the progress of the war. Then, with nothing else for him to do Cook read a book until lights out.

A short time before zero eight hundred hours, Cook opened the door to the guardhouse. With a MP brassard on the left arm, a sidearm in its holster hanging from a white web belt on his right hip, his musette bag bouncing on his left hip, and a roll of bedding under his left arm. "Hello, Sergeant. What do you want me to do with this bedding?"

"Put it on that table over there. You're Cook? Your man, Church, is almost ready to go, but you still have lots of time. Your train to London will not be arriving in the station for about an hour, but why don't you go ahead and sign for Church here, and for his personal effects here."

After he signed the forms Cook put the envelope containing the personal effects and his copies of the forms into his bag. "Okay, I'm already to go any time that someone can haul us to the railroad station."

---

[68] AWOL was usually articulated "a-wall."
[69] In fact pilfering from baggage by those who were responsible for its handling was common-place through out all branches of the services and in all war zones.

MILTON COOK

The jeep pulled up and stopped in front of the station. After pushing the passenger seat forward, he climbed out and turned around. "Okay, Church, you can get out now. I'll be right behind you. We are going into the station now. I have to get our tickets, and then we'll find some place to sit until our train gets here."

They located a partially occupied compartment with two side-by-side empty spaces, Cook sitting to the right of Church. Without much else to do, they watched out of the windows until the train slowed as it pulled into the station after about an hour's trip.

The train came to a complete stop in Liverpool station. The door was opened and the other passengers got up and started leaving the compartment, but Cook and Church remained sitting until they were alone. "Okay, Church. You stay in here until I get out, and then you can come out."

"We'll just hold up here for a little while until the crowd thins our a little."

"Okay, let's go. We are going down to the Underground and take a short ride to Euston Station."

Arriving at the Underground Station, Cook located and then dropped some coins into a vending machine to buy them two tickets to Euston Station. "Here, Church, take this ticket, and when you go through the turnstile, stop until I come through right behind you."

"Now, when we get onto a car, we will remain standing near the doors. It's only a short trip, four stations. Then, when we get off, there will probably be a crowd waiting to get on, so I will be right behind you holding on."

After getting off of the car, they ascended to the station concourse.

"This is our boarding gate for the train. So, we'll just park it here on this bench until they announce our train, and then we can go and get on board."

Moving with the crowd, they rapidly got on board one of the first of the railroad coaches standing alongside of the platform. "Okay, keep moving. I'm right behind you. Look for a compartment which still has two empty seats which are side by side."

Shortly after they sat down, the train started to move. Pulling out of the station, they were soon riding along past brick walls, over and under bridges, and past warehouses and all sorts of buildings and homes. The next three hours were devoted to watching the countryside and towns moving past, inspecting the few stations in which the train stopped, and enduring the coal smoke when it went through the tunnel.

"We are coming into Stafford now. We are getting off here and waiting for the local train that we will take to Stone. We have about a half-hour's layover in Stafford."

The train rumbled to a stop in Stone. Cook was up, opened the door, stepped out onto the platform and turned around. "Okay, Church, let's go. We will go over the tracks on that pedestrian bridge and then head down the sidewalk to that large tent. That's Stone Baggage where I'll arrange for us to get a ride out to Howard Hall."

"Hello, I'm Cook, I see that you don't have a truck here right now, so could you call the guardhouse and have them send someone down here to pick us up?"

"Okay," he said and picked up the phone and dialed the station. "Someone will be here to pick you two up in about fifteen minutes."

"Thanks for the call. We'll just stand over here and wait until our ride gets here."

Ogden slowed the jeep down, turned it and pulled through the gateway. "Hi, Cookie. You guys just got back? Well, the Desk Sergeant sent me down here to pick you up."

"Hi, Ogden. Thanks for getting here so fast."

Pulling the passenger seat forward, "you get in back, Church, and I'm right in beside you."

"Okay, Ogden, I'm in. Let's go."

The jeep backed out and around. Ogden shifted to forward and started down the hill. "How was London, Cookie?"

"I don't know. I was just passing through. All I saw of the place was two railroad stations and a short stretch of the Underground. I went down to an Airfield out of London to pick up this GI and bring him back here."

As soon as the jeep was parked by the guardhouse, Cook pushed the seat forward and climbed out. "Okay, Church, end of the line. You can get out now, and we will go inside."

"Hello, Sergeant, I'm back. This is Private Church, and here are his things. He's all yours now."

"Okay, Cook, that's good."

"Ogden, get Church a set of the special fatigues, and put him in the cell where he can change into them."

"Do you have any expenses to report, Cook?"

"The only thing that I paid for out of my pocket was for three short trips on the Underground, and they are not worth the trouble of the paperwork."

"Okay, good. I see by the duty roster that you are scheduled for patrol duty tomorrow night, but you have the rest of the day to do whatever you want to as you are off duty now."

Heading up the sidewalk on his way back to the barrack, Cook thought, *"After I get back to my room, park my things, change my uniform and clean up, it will be almost time to head to the mess hall for chow. I think that I might go see the movie tonight. Now that I'm back here and it's over with, that was a strange kind of trip. I went down to London, but I really wasn't there at all."*

CHAPTER 56

Tales from the Barrack-Company Picture

There had been a picture taken of the assembled 1257th MP Company in late July or Early August 1943, when most of our training had been all but completed. Now, about a year later, a second picture of the assembled company was scheduled. In order to photograph the entire 1257th, some other soldiers, members of the Repple Depple permanent party, would have to man the gates and guardhouse for the short time while the picture was taken.

On this overcast, but otherwise fairly nice day, we all were instructed to put on our duty uniforms. This meant uniforms pressed, brass and shoes shined, MP brassard, white leggings, white web belt with sidearm, white helmet liner and, for this occasion, white gloves. Leaving the barrack through the normally unused rear doors at the south end of the hallways, we proceeded past the back of the theater building towards Duncan Hall. Passing through a gap in the hedgerow, we entered into a field of deep green grass. This meadow was presently fallow farm field as for now it was with in the "confines" of this AAF Station.

A line of benches had been set up along the hedgerow, which ran to the east along the rear of the barrack. This hedgerow, through which we had just passed, was the nominal edge of this part of Howard Hall. The first thing was to form up I three files according to height. The tallest rank climbed up on the benches. The middle rank stood on the grass in front of the benches with the company officer in the exact center. Those in the shortest rank squatted more than knelt with the left leg forward and the right leg back. No one in the first row wanted to touch the wet grass with his knee.

As soon as the picture taking was completed, we were dismissed. Those who had duty assignment at the gates and guardhouse quickly departed. Those of us who remained were milling around slowly making our way back to the barrack in a kind of a Holiday mood. When suddenly without any warning, we were buzzed by a P-47 Thunderbolt. But we were cool and stood our ground to watch as this fighter pilot failed to intimidate this bunch of MPs. We stopped walking and watched as the airplane dived steeply and then pulled up about five feet from the grass. As the airplane turned and started to climb back up, small contrails formed at the tips of its wings, creating two white arcs to mark its path. This was kind of exciting, not something which we might normally expect to see even in our most unusual and varied duties.

1257 MILITARY POLICE COMPANY (AVN)

Buzzing was not a completely unusual occurrence. When they were in the area, the pilots would frequently fly over the station at a low altitude. There was one occasion when an attack bomber made a fast buzzing pass over the Repple Depple and then crashed and burned just north of Beatty Hall. But the P-47 was the one fighter airplane, which could dive steeply and pull out sharply just above the deck. This notable characteristic of the Thunderbolt would be the last thing that some enemy fighter pilots in hot pursuit would ever learn.

Chapter 57

Venerable Edifices Then and Again-Westminster Abbey

    Already this afternoon I had viewed Big Ben and the Houses of Parliament from the London Bridge, and then I walked on past them along an almost deserted street. This was the way to Westminster Abbey, and I was somewhat elated, as this was one of the places that I really wanted to take in. Approaching the entrance, I was greeted by a lone Englishman who offered his service as a guide for a tour through the Abbey. The fee, for which he would be my escort, was I thought modest and so agreed to his offer. I did think that it certainly would be a much more interesting tour with his commentary than it would have been going through the Abbey by myself. So I accepted his offer, and we were off for a jaunt through Westminster Abbey.

    As I recall, my guide must have been in his late thirties, at least that's how old he appeared to me from the vantage—point of my own youth. Although at the time, it did seem somewhat strange that he was not in the British Armed Forces or that he was not working in some war-related industry. It did occur to me that maybe he did have some sort of disability, and then again he just may have been working at night on some war-related job. So maybe working as a guide during the day, he was just supplementing his income off the record. Then again, maybe this was how he made his living—hiring himself out as a guide instead of working at a regular job. In any case, I did not think that it would be appropriate to ask him about these things. But then, of course, it may have been that he was in fact a German agent who was engaged in gleaning military information from the unwary GIs who took him as a guide. I kept wondering but did not think that a direct question was appropriate; as I did not expect a truth full answer in almost any case.

My escort did know his way around and through the "Abbey", it's history and all of the important places in the edifice. It was compelling to see all of the epitaphs carved in the stone markers under which rested the poets, the authors and the scientists of whom I knew. Each and every one of the stained glass windows had been removed and replaced with plain glass windows at the beginning of the war. The bland windows combined with the limited and subdued internal lighting inside the church created an overall gray almost dingy and gloomy aspect.

When we had concluded this very personal perusal of the places of interest inside, my guide then led the way out through a door to a place behind the edifice. Once there, I was amazed to see which were and had been some of the original Abbey workshops and living quarters. Some of these outbuildings had been damaged, and a few had been destroyed, during the Battle of Britain. Was it luck or German precision bombing which had spared the Abbey proper from a direct hit? One thing that really intrigued me was in one of the walkways running between the outbuildings. On the surface of a very large flagstone, an epitaph of sorts had been carved. The inscription in this slab proclaimed that beneath it rested the remains of, as I recall, about twelve monks. This group of monks had all succumbed at the same time to the plague, the Black Death, at some date during the Middle-Ages.

After completion of the outside tour, we went back through and out of the Abbey. Standing in front of the edifice as we were about to part-company, my guide recommended, what to him at least, was a good meal. He told me that I should go to this certain place that served this excellent steak and kidney pie. He further noted that this outstanding steak and kidney pie was served in a basin, a feature that he emphasized. This suggestion gave me pause for two reasons. First, I had never eaten any steak and kidney pie and did not think that I really wanted to try it. Steak and kidney pie was just not my cup of tea. Second, while I was more or less intellectually aware that when he said a "basin," what he was actually talking about was what I would call a "bowl." But nevertheless, I had this mental image of eating steak and kidney pie out of a washbasin. An eighteen inch diameter wash basin with a half inch wide rolled edge rim, six inches deep with curved sides and a slightly raised circular bottom inside of a half inch wide recessed ring. That was what I had in my mind.

The washbasin was used before and when sinks were not available. There was a hole through the rim from which the washbasin could be hung after the soapy and dirty water had been poured out. Maybe some people would eat out of a washbasin, but not me.

00000

It was in May of 1997, when my wife, Roberta, and I journeyed to England on a vacation. One of the "famous landmarks" included in the itinerary of the London portion of our packaged tour was perforce Westminster Abbey. The driver was just able to cram our tour bus into a small space among the myriad of already parked tour buses. After alighting from our tour bus, the very first thing to do was to somehow get across the very busy street and then carefully walk down the ancient appearing rough paving stone surfaced ramp leading down to the subterranean loos. With this group activity completed, we then crossed back over the busy street to our next stop on the tour, which was the souvenir and book store located in a building close by the entrance to the Abbey, to which it bore a likeness. Once inside the shop, we joined the milling throng scrutinizing all of the souvenirs and books. Having just the past weekend viewed a story on television about Lady Jane Gray, I took advantage of the many books on the history of the British Royalty to research her brief "reign." And so at last it was now our turn to stream through the Abbey. Once our tour group had assembled at the Abbey entrance, we were informed of the time to be back on board the bus. Turned loose, we were off straggling through, up around, down and back, taking in everything that could be viewed during the time allotted to pass through the edifice. In my mind's eye, I was startled by how bright and clear the interior appeared compared to the recollection of my prior visit. The reinstalled stained glass windows were imposing and the new window celebrating the service of the RAF in World War II was magnificent.

I was not surprised, however, that there was not even a hint of a tour of the outbuildings or of the fact that they even existed. Probably, I alone among the multitude of tourists on this day had special knowledge of just what lay behind the wall of Westminster Abbey. But still I knew that these outbuildings were there because I had seen them in a far different time in a far different world.

Chapter 58

Venerable Edifices Then and Again-The Tower of London

There was one and only one tour each day to view the inner tower of London. This visit was limited to—only those who were members of the Allied Military Services. Those who were eligible were required to enroll in the morning for admittance on that afternoon. It was mid October in 1945, when Szucs and I had come down to London on a forty-eight hour pass, that I first visited the tower.

All of the members of the various armed forces who had signed up that morning assembled at the entrance to the Tower of London before the tour starting time. Our gathering was greeted by several Beefeaters in their historical Tudor uniforms, and then separated into groups. Each band of military personnel set out following their particular Beefeater, as he led a circuit of the tower. At each of the many locations of the so very many historical happenings, my group would stop to hear the Beefeater's account of what and when and how each had occurred. The guided portion of the tour ended in a small Chapel, which we entered through a doorway located in a wall of the Great Hall. After this last visit on the guided tour, we were free to inspect all of the exhibits that we had just passed on the way going to the Chapel. This very large room was filled with row on row of glass display cases containing countless historical artifacts. One thing that I found fascinating was the extensive collection of gorgets. These gorgets were narrow, crescent-shaped pieces of shining metal that had a small link chain attached between both of the pointed ends. At one time, these ornate bits of gleaming metal had been worn at the neck of officers to denote that they were on duty. It occurred to me at the time that the gorget was the very last bit of a degenerated breastplate—the final remnant of the knight in shining armor.

Passing back and forth between the glass museum-cases, viewing the collections of artifacts, I came back to the Chapel entrance at the end of a row. On the spur of the moment, I reentered this anteroom to see it one more time. There I was all-alone in this small room with its arched roof and half dome over the pulpit. At least I was all-alone in the present time, but not in the past, to which I was somehow inexplicably linked to long gone people and long past events. Somehow, I became aware of the past, a sense of things that long ago had taken place in this room. It was transcendent. It was *je ne sais quoi*.

00000

In July of 1949, I was back in London, literally waiting for my ship to come in. Having completed training at the U.S. Naval School for Advanced Undersea Weapons in Key West, Florida, I was traveling under orders to rejoin my ship, the U.S.S. O'Hare DD889. Until she made port in Plymouth, England, I was attached to the American Embassy on Grosvenor Square. During the short layover, I was billeted with the enlisted men, Marines and Navy corpsmen, who were stationed at the Embassy. While waiting for my ship to come in, I was assigned temporary duty working in the Embassy lunchroom.[70] The lunchroom was not open on the weekend, so not having anything special to do, I thought why not visit the Tower of London again. At this time, the Tower was again open to the general public and was once more a major tourist attraction. There were many groups of sightseers making the circuit, a very great crowd when compared to the scattered few on my first visit to the Tower, but still the number of visitors on this day was more than likely much, much smaller than the throngs of future times. Evading any conducted tour, I proceeded directly to the great room filled with the rows of museum display cases. Looking over the exhibits in the glass cases once again, I worked my way towards the Chapel doorway.

---

[70] At this time, I was a Petty Officer, Second Class (E 5) electronics technician, and it was the first and only occasion while in the U.S. Navy, on which I had ever been assigned any job which remotely resembled K.P. duty, but that's another story.

Biding my time, after a short wait, the Chapel was empty. Now was my chance so quickly, I entered this small chamber. But on this visit, it was just another room, and there was not even a hint of a connection to the past. Maybe it was just too many tourists that had disordered the link to the past. Then again, maybe, just maybe, what I had experienced before in this Chapel had only been in my mind. This was very disappointing. I really believed that I could re-experience the unusual occurrence of my first solo time in this room. But I suppose that if I did have the experience that I remembered had occurred at all, it was a once in a lifetime event.

Chapter 59

Singular Service-Orders to Paris

The MP patrol jeep rolled through the Howard Hall Gate, pulled over and came to a stop. The MP sitting on the right pivoted, swung out and around, moving quickly towards the front of the vehicle. The seat that he just abruptly vacated banged forward, and the four MPs who were returning from patrol duty in Eccleshall and Millmeese spilled out. Forming up in a single file, they hastily entered the guardhouse and headed for the latrine. Then, one by one, they made their way out through the doorway and on around the guardhouse to the mess room in the rear corner. On the way out, as they passed the desk sergeant, he said to Foiles and Cook, "After you have finished your snack, come back in here and see me. There is a very special assignment that I need to talk to you about."

It was typically nippy on this March night, and the metal benches of the picnic table were just too darned cold to sit down on. That being the case, they drank their coffee and ate the fried ham sandwiches while standing around and talking. "Well, Foiles, I wonder what the heck they are going to have us doing now."

"Well, let's go back inside and find out, Cookie." They sat the empty coffee mugs on the counter top of the mess room half door and walked back around the two corners and into the guardhouse.

"Foiles, you and Cook will be flying to Paris tomorrow morning. You will be escorting two soldiers back here, who are being held in custody at the Repple Depple facility there. Foiles, these are your orders. You and Cook are to be back here ready to go at zero eight hundred hours. Then, someone will transport you to the airport."

Tomorrow was Saturday, and neither Foiles nor Cook had a duty assignment. They might have gone some place on a pass if they had chosen to do so. However, they were both good soldiers, and so they did not complain at all about being sent on a special assignment to Paris. They resolved to do this extra duty assignment, as ordered, to the best of their ability.

After they had returned to the barrack, Foiles said, "Here, Cookie, you have been going on these trips and know what is going on, so why don't you carry the orders like you usually do."

"Sure, Foiles," Cook said as he took the orders from Foiles.

Quickly reading the orders, he said, "I see that these orders are five days, which will give us a lot of time in Paris, because it will only take a half of a day each to go there and come back. So be sure to put a six-day supply of underwear and socks in your bag. See you in the morning."

His roommate had also just returned from patrol duty and was still getting into his bunk. So, Cook took time to read the orders in detail before he turned out the light and hit the sack.

Sgt. Foiles and Pfc. Cook on temporary duty not to exceed five days to go to Paris, France for the purpose of escorting Technician 5th Grade Edward E. Webster and Staff Sergeant Barney E. Cash in confinement at AAF Station 397 to confinement at AAF Station 652, which was here. The multiple copies of Special Orders No. 82, Dated 23 Mar 1945, had been typed on a light-weight, gray, eight by ten and a half inch paper. Each sheet of the orders had been crisply embossed with the H.Q. 70th Reinforcement Depot seal to render them official.

After laying the set of orders on the top of the built-in drawers, he removed his uniform, hung it up in the locker, turned out the light and got into his bunk.

A short while before zero eight hundred hours Foiles and Cook entered the guardhouse. "Good morning, Sergeant. Here we are, ready to go off to see Paris."

"Good morning, Foiles and Cook. You two go on out and get into the jeep. Maddox will drive you over to Seighford Airfield."

A short ways to the west of Stafford, Seighford Airfield was about eight and a half miles from the Repple Depple. Driving on the roads through the countryside, the trip to the airport took a little less than a half an hour. When they were logged in at the gate, the MP provided directions to the flight operations. Maddox stopped the jeep in front of the building, "Okay, here we are. You two have fun in Paris. See you later."

"Okay, Maddox. Thanks for the ride. And yes, we do intend to have some fun while in Paris."

"Let's go on in here now Cookie, and find out just what we have to do to catch our flight."

Seighford Airfield flight operations was located in a gray, single story building with walls of the typical asbestos reinforced concrete sheeting. The entrance was centered between two large windows. Identical windows ran along both of the side-walls. A small porch like step with large gaps between the boards protruded from just under the door. Entering through the doorway, they were in an area about six feet wide that ran across the building. This area was separated from the rest of the large open room by a closed front, three-foot high counter. Tables, chairs, desks with the typewriters and telephones, filing cabinets and a Teletype console were arrayed in the larger part of the building behind the counter. Several "EMs" and officers were working behind some of the desks.

Foiles stopped in front of the Staff Sergeant who was standing behind the center of the counter. "Hello, Sergeant. We are checking in. We are to fly to Paris this morning."

"Okay, let me see your orders."

The Sergeant quickly scanned the orders. "Oh yeah, here you two are on the manifest for today's flight to Paris on the Seventh Reinforcement Depot's C-47. You two are here really early. That aircraft won't be departing from here for another two or three hours. You go on outside and hang around. We will announce it over the P.A when it's time for you to get on board your aircraft."

Cook picked up their orders, folded them up and then placed them in one of the inside pockets of his Eisenhower jacket. The inside pockets in the Ike jackets were located higher and were somewhat shorter than those in the older service jacket that it had replaced.

So here they were with nothing to do and two or three hours to kill. This was Army SOP, the old hurry up and wait routine, nothing new here. There were some wooden benches to sit on. These benches became harder and harder as time passed. So after a while, they decided to get up and wander around with all of the other soldiers who were also waiting to be called to board their flights. There was even a Brigadier General waiting to get back to Europe, who also was milling around just like the rest of the GIs who were also waiting to catch a flight there.

Finally, at last, they were called. Now it was time to get on board the airplane for their flight to Paris. A soldier checked their names on the manifest and then directed them to the aircraft. Walking out on the flight line, they saw that this particular C-47 had been named. Painted below the side cockpit window there was, the name "Froggy Bottom Express" and a picture of a green bullfrog. The name of this plane was probably a reference to the environment at the location of this airfield. This aircraft and crew were assigned the Repple Depple and provided the courier service to the station in Paris where they were headed.

Holding onto the side of the cargo doorway, they used a small ladder, which was inserted below the door to climb up into the fuselage. Then they walked up the sloping deck towards the front of the aircraft, passing by the cargo. The packages had been lashed to the floor using the rings that were flipped up from their normally recessed locations in the floor. Picking two seats, rectangular indentations in the aluminum benches that were hinged to both sides of the fuselage, they sat down. These aluminum bench seats in cargo aircraft were true bucket seats. One of the aircraft's crew told them to fasten their seat belts as he headed back to close and latch the cargo doors. After closing the doors, he came back and disappeared into the crew compartment. A starter screamed up, and the engine sputtered, snorted, coughed, barked and started with a roar. Then the second engine repeated the same routine; except now it was much less noticeably as, its start-up, was masked by the rumble of the already running engine.

Foiles and Cook had been in the Army Air Force for over two years, and now at last, they were finally going to be flying. In fact, this would be the first time either one of them had ever been off of the ground in an airplane.

The C-47 taxied to the end of the runway, turned, revved up its engines, released its brakes and started accelerating up the runway. The tail wheel lifted off of the pavement, and for a short while the cargo floor was level. Then the aircraft was in the air climbing, and the floor sloped even more than it had when parked on the ground. When the Skytrain reached its cruising altitude, the deck was once again level. Heading almost due south at 170 miles per hour, they would be landing in Paris in a little over two hours.

The sky over Europe was free of German aircraft. The Allied Air Forces were maintaining standing patrols, day and night, over all of the Luftwaffe airfields to keep it grounded. On the second of March, the U.S. Ninth Army had captured the almost intact Remagen Bridge, and was advancing across the Rhine in that area. Today was the twenty-fourth, and while Foiles and Cook were not yet aware of it, just the night before General Patton's U.S. 12th Army group had crossed the Rhine south of Mainz. Then last night, the British XII and XXX Corps and the U.S. XVI Corp had crossed the Rhine on both sides of the Lidde River. The tide of battle had crossed the Rhine and was flowing into Germany now.

Reaching inside his Ike jacket to retrieve their orders, Cook's hand found nothing. "Oh crap, Foiles, I've lost our orders."

"Are you sure?"

"Oh yeah, I'm sure."

"That's terrible! What are we going to do?"

"Well, it's really dumb, and its very embarrassing, but I don't think that it will be a real big problem. You remember what they told us back in basic training. If you really mess up and do something wrong, you should come forward and disclose it. Then the Army will not hold it against you. So I'll just have to report to the Provost Marshal when we get there and tell him that I carelessly lost our orders. That should take care of losing our copies. The already have our orders by Teletype. That's how they know that we are coming to Paris. So I'll just have to suggest—request—that he have some copies of the orders typed up from the Teletype message for us."

"Well, that sounds like it will work. We'll just have to wait and see if this all turns out okay like you say."

After passing over the southern coast of England, the aircraft turned east and before long was crossing the western coast of France. Then a short time later it started to let down to make the landing at Paris. Once on the ground, the airplane turned off from one of the Le Bourget Airport runways and taxied over to the U.S. Army Air Force flight operations area. Shortly after the cargo doors were opened, they scrambled down onto the concrete surface of the flight line. A GI who was standing at the side of a weapons carrier, into which the cargo from the C-47 was being loaded said, "Are you two Foiles and Cook?"

"Yes, we are."

"Okay, I thought so. You go ahead and get into the front seat. As soon as all of these boxes are loaded in the truck, we will be on our way to the Repple Depple."

"We're MPs who are stationed at the Repple Depple headquarters at Yarnfield. They sent us over here on five days' TDY to escort two AWOL GIs back there."

"That sounds like a good deal to me. I'm sure that you two will have a good time while you're here."

The driver let out the clutch, and they were on the way to that area of Paris that was known as the *Bois DeBologne*, the Chateau Rothshield to be specific, a large estate backed up to the Seine; the location of the Repple Depple in Paris. Most of the personnel processed at this station were assigned to the Ninth Air Force.

Arriving at the Chateau, the driver stopped the vehicle at the gate to be logged in. Foiles said to the MP, who was entering the information on the log sheet, "Where is your guardhouse? We have to report in."

Entering the room that they had been directed to, which served as the guardhouse office, they found the desk sergeant sitting behind a table directly in front of the doorway. "Good afternoon Sergeant, we are Sergeant Foiles and PFC Cook. We are here to take a couple of GIs off of your hands in a few days."

"Good afternoon, Foiles and Cook. We have been expecting you, and have two billets assigned for your use. Corporal O'Sullivan will show you where your bunks are located, fix you up with some bedding and clue you in on the mess hall and latrine."

"Thanks, Sergeant. That's good, but we have a serious problem and need to report to the Provost Marshal first. It's a matter of lost orders."

"Okay, okay. I'll inform him that you two need to see him right now."

Upon entering the office of the Provost Marshal, they halted in front of his desk, came to attention and saluted. "Sergeant Foiles and PFC Cook requesting permission to speak to you, Sir."

Returning their salute, he said, "At ease, soldiers. The Sergeant said that you have a problem that you wanted to talk to me about."

"Yes, Sir. I am PFC Cook and I need to report that I carelessly lost our orders while we were waiting for our flight from the airport in England. We were there for a very long time before we could get on the airplane and fly over here."

"Well, you are correct about being careless Cook. What do you think that we should do about your orders now?"

"Well, Sir, I know that our orders are already here by Teletype. So if you could have one of the clerks type some carbon copies for us and the original for you, and then sign them, I think that should take care of the problem, Sir."

"That sounds alright to me. I'll have some copies of your orders typed, and you should be more careful in the future. You are both dismissed." They came to attention, saluted, executed an about-face and then quickly exited from his office.

The Provost Marshal may have thought that it was very unusual that the Private instead of the Sergeant had been carrying the orders and had also been the one to report them lost. However, he did not make any comment about who had the orders or who had reported them lost.

As soon as they were out of the Provost Marshal's office, Foiles said, "Well that sure worked out okay, just like you said it would. Now, O'Sullivan, we are ready to go with you. And Sergeant, the Provost Marshal said that we could have some copies of our orders typed up. Can you arrange to have someone type us some copies. We will need a copy of the orders to get off of this station and to return to England."

"Alright, Foiles, I'll take care of getting copies of your orders typed up for you. You can go on with O'Sullivan while he gets you some bedding from the guardroom stock, and then shows you where everything is located. After you are all through with finding your way around, you can come back here and maybe the copies of your orders will be ready. You two sure pulled yourselves out of that one."

O'Sullivan handed Foiles and Cook each a mattress cover and two OD blankets. "Okay, come with me. Your bunks are in a room up these stairs."

"Here we are. Just put your things on whichever of these two bunks each of you want."

Foiles and Cook each selected a bunk and sat the bedding and their bags on top of the bare pad.

"Now we will go back down the stairs to pay a visit to the latrine. You probably want to go there now anyway."

Leaving the Chateau proper through a side door, they saw a group of pyramidal type tents. The tents were standing in the area that before the war had been a formal garden. The latrine, with its duckboard floors, was situated in these tents. In some of the tents wash basins and showers with hot and cold running water was available. In separate tents were the toilets, field latrines with long wooden boxes supported over long deep trenches. Round holes of the appropriate size were located strategically along the top surface of the box.

As the were heading back into the Chateau, Foiles said, "Before we go to the mess hall, we need to change our money from English to French, and then we should check back in the guardroom and see if the copies of our orders are ready."

CHATEAU ROTHSCHILD

In the foreground the mess tent is erected on the entrance terrace. Note the chimney pipe from the mess tent stoves that runs up the out side of the chateau and then above the top of the roof.

"Yes, that's a good idea, Foiles. Let's go to financial first so that you can exchange your money and then stop by the guardroom to check on your orders."

With their money exchanged and the fresh copies of orders tucked away, it was time to visit the most important place, the mess hall, which was in fact a large Army mess tent. The tent had been erected on a terrace on the side of the Chateau opposite of the latrine in the garden. With O'Sullivan leading the way, they grabbed mess trays and utensils after getting in line to make a pass through the chow line. While they were eating their supper, O'Sullivan suggested,

"Why don't you two meet me in front of the Chateau about eighteen hundred hours, and I'll take you down to the local bistro where we all hang out."

"Okay, that sounds like fun. We'll be out there ready to go. See you then."

The tall wrought iron bars of the fence bounding the street side of the Rothshield Estate stopped and made a turn through a monumental ornate stone column and then headed for the river. The soldier did not make a turn, but instead continued on in the direction that they were headed. Not too far beyond this point, they arrived at the very local bistro. After the obligatory introductions to all of the attending members of the local permanent party, Foiles and Cook joined the group for many, many bottles of cheap champagne. Even if it was the bottom of the line, this wine was fine, a welcome change from the warm British brew, which was their regular drink of "choice." This was indeed a fine way to start this fine, but short, duty assignment to the city of Paris.

In the morning, in the mess tent for breakfast, fried eggs and bacon, toast, coffee and fresh grapefruit were on the menu. Now all of the KPs were in fact German POWs. That's one of the fortunes of war, the POWs can get stuck with the kitchen police duty, which when you think about it is not a very bad job to have as a prisoner of war. At least they eat well. However, the grapefruit were all cut in half the wrong way, with instead of across the segments. This was a different experience eating grapefruit that had been cloven in this strange manner. These halves of grapefruit essentially had to be peeled so that they could be eaten. A grapefruit cut in two in this manner could not reasonably be called one of the fruits of victory. So here is the question: were these grapefruit cut in half-antipodal, out of malice or because of stupidity? Were these German POWs sore losers exacting a bit of revenge for their present situation by rendering the grapefruit somewhat difficult to eat. Or were these German POWs so unfamiliar with grapefruit that they had no idea at all of the proper and civilized method to slice and serve this citrus fruit. If they were not aware of the correct method of serving grapefruit then there was no comprehensible way that the Germans could have ever been deemed "the master race."

Stopping by the EM lounge before departing from the station on the first full day of their tour, Foiles and Cook each picked up an *American Red Cross Guide to Paris*. A map of the Paris Metro, with a list providing the locations of all of the military clubs, was centered above a slightly larger map of the central portion of the city. On either side of these maps, there were lists giving the locations of public administration buildings, railroad stations, academic buildings, parks and sports and music halls. A second list of places to see identified and provided the location of the main buildings, museums and religious services. Centered below the map of central Paris, there was a very short course on how to ask your way with questions and answers in English and in French, with pronunciation guides. Below this little lesson were some general notes with the last being "this map may be mailed home." The American Red Cross guide to Paris had been printed on the reverse side of a strip from a German ordinance map. On the back of Cook's ARC guide was a map of a small area of Ireland, west of Dundalk that included area north and east of Carrick Macross. This was in fact a British ordinance survey map (which was still noted in original English caption) with legends in German and metric scales on the upper and the lower borders. This German ordinance map was most probably residue of the aborted "Operation Sea Lion." These precise and very detailed British ordinance survey maps would have proved to be very helpful to the Germans, had they acutely invaded the British Isles.

The German Army had occupied Paris, which had been declared an open city, on June 14, 1940. The city had, in turn been liberated by the French and the American Armies on August 25, 1944, just seven months before to the day. Now, on this morning, the Allied Armies were engaged in battle with the German Armies ranging 200 to 300 miles to the east. But here, in Paris, it was clean and bright without any obvious signs of war's destruction, or of the German occupation for that matter. Paris was a very pronounced contrast to the dull gray war weary Britain, pocked as it was with archipelagos of bombed out structures, from which they had departed just the day before. Members of all of the diverse services of all of the various Allies were seen everywhere much as they had been in London before the invasion of Europe.

Going to wherever one wanted to go in Paris was very easy using the Metro Subway system. It was also very cheap because the Metro was in fact free for all of the Allied military personnel at this time. But most of Foiles and Cook's tour was spent just walking around the city to take in all of the sights. They visited many well-known locales such as *des Champs Elysees,* the even wider *Avenue Foch,* the *Arc de Triompe,* Notre Dame and the Eiffel Tower (which was closed at the time). And of course, they had to see *Place Pigalle,* which was otherwise known far and wide by the American soldiers as "Pig Alley." On one day's tour, they came across large posters at the front of a theater, the Olimpia Hall, which advertised concerts to be performed by the American Band of the Allied Expeditionary Force. This band, until a short time before, had been directed and commanded by Major Glenn Miller. Major Miller had vanished on 15 December 1944, while piloting a small aircraft across the English Channel. The disappearance of Major Miller is one of the enduring mysteries of World War II.

On yet another day's tour while walking along the avenue, they came upon some unusual, at this time, portrait photos in the display window of a photography studio. The entire surfaces of the photos were covered with very small rounded horizontal ribs. These small flutes somehow resulted in the creation of a three-dimensional appearance on the face in the photo.

These two soldiers knew very well that being in Paris, they must do all of the tourist things, which included shopping. Cook and Foiles went into many, many shops and stores looking for things that would make fine gifts from Paris. Gifts that would be sent back home to the United States just as soon as they returned to England. Cook also purchased some special victory supplies.

During all of the rambling around and through Paris, they came across only a few clean, empty lots and no damaged buildings at all. The absence of destruction was a sharp contrast to the large scale, still standing devastation seen in the larger, and some of the smaller, cosmopolitan areas of England.

At the end of each day's roving about the city sightseeing and shopping, these two GIs managed to be back at the Chateau in time for chow. After eating supper, as a matter of course, they would join the stroll down to the local bistro, where seated around tables in front of the bistro, they joined the group partaking of the cheap champagne for the remaining few days of temporary duty in Paris. This Paris experience was all so different from what was available for entertainment in Britain.

PARIS 1945

Foiles and Cook doing the tourist thing, having their picture taken before the Eiffel Tower

First thing after morning chow on Wednesday, the twenty-eighth, before starting out on the last day of sightseeing and shopping, Foiles and Cook reported to the guardroom. "Good morning, Sergeant. I'm Sergeant Foiles, and this is PFC Cook. We need to know what time to be back here tomorrow morning to claim custody of the two GIs who we will be escorting back to England. Also, we need to find out about transportation to the airport."

"Good Morning, Sergeant. It's good that you two checked in here early this morning. Now we won't have to send someone looking for you two. You need to be here in the guardroom at zero nine hundred hours. Your prisoners will be ready to go at that time. A truck will be ready that will transport the four of you to the Airport. The truck will depart as soon as it has been loaded with whatever is being shipped to England in the aircraft with you. The driver will drop you off at the security control office for clearance before you will be able to get on board your flight back to England. Any questions? No? Have you been having fun here in Paris?"

"Yes we have! This is all so different from where we are stationed at the Repple Depple in England. Except for a seven-day furlough that we had in England, this is the longest time that we have not had any real duty for the entire time that we have been in the Army. We will be back here at zero nine hundred hours tomorrow ready to go."

This time when entering the guardroom, they were dressed for duty with MP brassards on the left sleeve, side arm on the right hip and musette bag hanging at the left hip. "Good morning, Sergeant. Foiles and Cook, reporting for duty. We are ready to take custody of the two soldiers who we are to escort to England today."

"Good morning to you also. Okay, Foiles, you sign here for the custody of Staff Sergeant Cash, and here for his personal effects in this envelope. That's your man sitting in the left chair against the wall."

"Okay, Cook, now you sign here for the custody of Tech Five Webster, and here for his personal effects that are in this envelope. Your man is the GI sitting on the right. There is a weapons—carrier parked out in front of the Chateau that's being loaded right now. So you can all go out now and get into the rear of the truck. They probably are still loading the cargo that is being transported to England today. So you will have to work your way around whatever will be in the truck."

"Okay, you two on your feet and out the door. Cook and I will be right behind you!"

When they reached the bottom of the steps, Foiles asked the soldier standing by the weapons carrier, "Is this the vehicle that we're to ride in for the trip to the Le Bourget airfield?"

"Yep, this is it. You guys go ahead and get up in the back. I made sure that both benches were clear to sit on. As soon as one more box is loaded, we'll be on our way."

Foiles ordered the two prisoners to get into the truck and sit on the left side. And then Cook and Foiles climbed up into the truck bed and sat down on the right side bench. The last box was then loaded, and the driver closed and secured the tailgate. Then he went around, climbed in the cab, started the engine and they were on the way for the short trip to the airport.

Cook was familiar with some of what he thought might be part of the cargo today. Sometime before, one of the mess hall cooks, who had been one of the MP company cooks, had given him several slices of a baguette that had been left over from the permanent party commissioned officers' mess. At the time, he had wondered just how that bread had gotten from France to England, and now he knew. That French bread had been like eating cake compared to the course gray multigrain English bread that was served in the "EM" mess hall.

While some part of this cargo had to be U.S. Army documents, there were just too many boxes to take care of official business. So, Cook speculated, with good reason, that besides French bread here in the cargo, there were rations for the permanent party officers' mess and club. Such things as French cheeses, wines and spirits were more than likely in some of the boxes. There was a real advantage of having an aircraft assigned for the exclusive use of the Repple Depple.

Airplane engines were heard rumbling in several different locations, close and far away, when the weapons carrier turned and went a short way before making an abrupt stop. The driver appeared at the rear of the truck and lowered the tailgate. "This is the security control office. You will have to check through it before you can go out onto the flight line and get on board your aircraft. As soon as you get out I've got to go on around so they can load all of this stuff onto the airplane."

"Thanks for the ride," Foiles said as he hopped down from the rear of the truck.

Cook jumped off right behind him, and as he hit the ground, they both turned around to look back into the vehicle. "Okay, you two come on. Get out now."

"Now go in through that door, we will be right behind you all of the way."

Both of the MPs laid the copies of their orders on the countertop. "Lieutenant, Sir, these two soldiers are in our custody. We are escorting them back to the 70th Reinforcement Depot in England."

"I see that you both have the same orders. Do either of you have any uncensored or unauthorized classified material in your position?"

"No, Sir!" they both replied.

"I did not think so."

The officer completed two security controls forms, using the information contained in the orders. Then he said as he was signing the second one, "Oops, wrong date."

So he crossed out the date that he had just stamped on the two forms. Then he set the date stamp to the correct date, March 29, 1945, and re-stamped the forms above the crossed out, incorrect date. "All right, Sergeant Foiles, sign the declaration that you do not have any uncensored or unauthorized classified material in your possession. And you, PFC Cook, sign this one here."

The lieutenant handed a signed form back to Foiles and Cook, "Okay, you show these to the traffic officer before you get on board your aircraft. Go on out the door over there to your left. Flight operations will be in the first building on your right."

With the two prisoners leading the way, they entered the flight operations building. Keeping the two soldiers between them, Foiles presented his orders to the Sergeant on duty behind the counter.

Reading the orders, he said, "Alright, your aircraft is almost ready for you to get on board it. Just as soon as all of the cargo is loaded and secured, they will let you get into it."

"Do you or someone else in here want to see these security forms?"

"You guys just went through security before you came in here didn't you? So why would anyone in here want to check them. Go on and get on board your plane."

No one else asked to see the security forms before they climbed in the C-47. No one ever asked to see or turn in the forms after they had arrived in England. After all they were MPs on duty, and who would ever think of checking the papers of military police under these circumstances.

The tires screeched and puffed smoke when they contacted the runway as the Skytrain landed at Seighford Airfield. The aircraft was taxiing to the Army Air Force operations area when Foiles remarked, "I wonder if there will be anyone here to pick us up now? Okay, Cash and Webster, both of you pay attention. Cook and I will get off of the aircraft first. Then you two will get off when we tell you to."

When they came around the corner of the flight operations building, Foiles' question was answered. Because there was Clancy, sitting behind the steering wheel of the M.P. jeep, waiting to pick them up.

"Oh crud! They knew that we would have two GIs in custody. I wonder why they did not send something bigger to pick us up?"

While it was not that unusual for six MPs to crowd into a jeep, this was just a slightly different situation.

"Hi, Clancy, you here to pick us up? How come you don't have a weapons carrier?"

"Hi, Foiles, Cookie. How was Paris? We tried, but could not get one on short notice. So they sent me over to pick you up in the jeep."

"Well okay. It's a short trip anyway. Cash, you and Webster get into the rear seat. Then Cook will get in with you two, and sit on the right side. Paris was fun, by the way."

After the three soldiers were jammed into the back seat, Foiles pulled the passenger seat back to its full upright position, hitting Cook's knees in the process, and then climbed into the jeep.

The vehicle rolled slowly through the Howard Hall Gate without stopping and parked alongside the guardhouse. Raines, who was on duty, said, "Ha! So at last the two of you are finally back here. How was Paris?"

"It was fun, Raines. Fun." Foiles exclaimed as he swung out of the jeep, turned around and flopped the seat forward.

Climbing out next, Cook turned to the two GIs still in the back seat, "Okay, you two can get out now. It's the end of the line. Now we'll just go around the corner and in through the door."

"Hello, Sergeant. Well, we are back. This is Cash, and this is Webster, who are both ready to go into the cell. Cash's personal effects are in this envelope."

"And here are Webster's things in this envelope," Cook said as he handed it to the desk sergeant.

"Well, did you two have a good time in Paris? We did not know just exactly when you two would return, so we couldn't schedule you for duty. So you have the rest of the day off. Foiles, you have the duty here in the guardhouse at zero eight hundred hours tomorrow. And Cook, you are assigned to the Duncan Hall exit at the same time."

"Okay, we'll be back down here in the morning. Let's go, Cookie." Exiting the guardhouse, they headed up the sidewalk on the way back to the barrack. On the way, they passed the mess hall, to which they would shortly be back as it was time for noon chow.

"You know, Foiles, aside from losing the orders, I think that this was the best duty assignment that we ever had or probably will ever have in this Army."

"I have to agree with you, Cookie. I'm sure that you're really right."

And after all they were good soldiers, who, as it has so often been said, were just following orders.

Chapter 60

A View From the Barrack-VE Day

The war in the European Theater was finally over. Hostilities had ended. President Truman had proclaimed the eighth of May 1945 as VE Day. The last German forces would not surrender until the twelfth of May, but all of the fighting had stopped on the eighth of May. The eighth of May 1945 was the day, which fixed the turning point—the very beginning of the long journey home.

But World War II was not over, as fighting still continued in the Pacific Theater of Operations, so VE Day was only it culmination in Europe. There were millions of troops in the ETO, most of which had to be returned back to the United States. While some of the armed forces must, for the time being take-up the occupation of Germany, the majority could now be returned to America. The first to leave would be those combat units that had seen battle in the ETO, and now would start training for the coming invasion of Japan. These harden war wise soldiers who had endured the invasion and combat across Europe would now, with their acquired expertise, get to do it all over again in Japan. For the rest of the soldiers in the Army, the order of return would be based on their point count. Points were awarded for the length of service time in the Army, the length of service time overseas, and the length of time in combat. Points were also conferred for combat decorations according to their order of precedence. So the more points a GI "had acquired", the sooner that GI would be leaving the ETO for home. The emphasis of the Reinforcement Depots had now shifted from processing GIs newly arrived from the United States to the processing of them for their repatriation back to the United States.

But still, the war in Europe was concluded. The Allies were victorious; Germany and Italy had been defeated. There will be those who say that the Allies had "won" the war, and therefore Germany had "lost" the war. But is this true? War is not a game. How would you score a war? Who officiates a war? Who keeps track of the points? No, it was a victory and a defeat, but there would be no grand inscribed trophy awarded for "winning" this war.

Proficiency badges are won because they are awarded based on a score against a standard, but medals are never won. Still, there will be some who, will say or write that someone has "won" a certain decoration. Medals are awarded the recipient based on accomplishment. Some awards are for service, for being there at a time or at a place and for doing your job responsibly. Some higher decorations are awarded by a citation for some meritorious service. The decorations stemming from battle are awarded to the recipient by a citation for heroism, extraordinary achievement involving combat and for military merit. But medals are never, ever won! It is highly unlikely that any soldier has ever said to him self, "I think that I will get myself wounded, so that I can "win" a Purple Heart."

CHAPTER 61

Occurrences in the Barrack-A Party of GIs

It was V.E. Day, Victory in Europe Day. The war in Europe was at an end. It was time to celebrate. This day, the day that so very many had waited so very long for, had at last arrived. It was time to celebrate. With the end of the war, we would now soon be going home. It was time to celebrate, and it was very fortuitous that on this very day, my platoon was the one, which was off duty, so we could all go to the town and join in the celebration.

But before it was time to take off and join in the celebration, we were stunned when the platoon was designated a standby unit. Now we would not be going anywhere on a pass. The Provost Marshal had made a decision that in the event of the celebrating getting out of hand and some disorder or in the event that a riot developed, we would be available to squelch the disturbance. This was just one more caprice of the fate of war. It was a very lucky break to have this very day free from duty when the hostilities ended, but then to be more of less confined to the quarters because we had this very day off.

It was highly improbable that a bunch of Army Air Force GIs could manage to get enough to drink, in the local pubs, that they would be roused to the point of getting out of hand. These pubs would be packed to over-capacity, wall-to-wall drinkers, and standing-room only. So it would be slow and difficult to just get hold of a pint of beer. Refills would be spaced out because of the time it would take just to get back to the bar. This was just as well as much the same situation would exist when after the alcohol had been extracted, the residue of the beer had to be disposed down a drain. Also, the many celebrating young women would divert the attention of most of the GIs who might have otherwise considered undertaking a riot. But nevertheless, here we were stuck in the barrack; and this was where we likely would stay for the rest of the night, unless callout for very unlikely, but no doubt exciting, duty.

The rest of the Company had departed off to their various posting, out of the barrack and at least being able to celebrate vicariously (at least they should be only celebrating vicariously). Those of us who had to remain behind were not a happy band of soldiers. It was time to go out and celebrate, and we were probably not going anywhere except into our bunks. Having to remain behind in the barrack when there was so much excitement everywhere else was an extreme frustration.

Although I never could have foreseen this present state of affairs, I was in fact prepared for our present situation. This was because when Foiles and I had returned from our TDY at Paris, I had stowed in my bag some victory supplies. As it turned out these were just the right provisions for this very special occasion. Now, it was time to retrieve the victory cache from my locker. Now that those of us who remained behind on standby were all alone, by ourselves, in our deep disappointment, it was time to make my move. Opening the door to my room, I stepped out into the hallway and yelled, "Okay, everyone out into the hallway, and bring your canteen cups."

Doors swung back and heads popped out, turning to look down the hallway. Everyone was wondering just what the hell I was up to now. "Alright, from Paris, I have one bottle of champagne to toast the victory, and one bottle of cognac to celebrate the victory."

Now that I had my comrade's attention, the quickly issued forth into the hall with canteen cups clutched in hand.

The warm bottle of cheap champagne was popped open and shared around into everyone's aluminum mug. Then this band of standbys drank a toast to the victory. With the champagne so quickly gone, it was now time to remove the stopper from the bottle of cheap cognac. Then, close as it could be judged, equal amounts of the amber liquid was meticulously rationed into each one in the confluence of canteen cups. Not being well versed in the consumption of spirits, it took us all a while to finish off our brandy. Now just a little bit loaded, we were probably not in the best of shape to put down a riot, but we sure as hell would have enjoyed doing it.

The celebratory drinking was completed; the cognac was all consumed, when someone near the end of the hall said, "I'm hungry. Let's go and get some chow."

"Hey, that's a good idea. Let's go over to the snack bar."

"To heck with that, I said chow, now a snack. I mean let's head over to Duncan Hall."

"Duncan Hall? Why would we go over to Duncan Hall to get some chow?"

"Because, as you obviously don't know, there is always a chow line set up in the mess hall there. They keep it open in case they have to feed some late arriving travelers."

"Oh! I didn't know that, but it's a good idea. Let's go on over there and get something to eat."

Then a chorus, "Let's go. Let's go get some chow."

So off we went, out of the barrack, a very happy band of MPs. Then someone started to sing and all joined it on one of our marching songs:

> *I've got sixpence,*
> *Jolly, jolly sixpence,*
> *I've got sixpence to last me all my life . . .*

And so it went all of the way down to the Howard Hall Gate.

Splann, on duty at the gate this day said, "Where did you guys get the booze?" And he knew well about booze, because in real life he had been a "professional" moonshiner.

"It's a military secret, and we cannot provide that information to you."

"Loose lips sink ships."

"What are you guys doing here? Where the heck are you going?"

"We are all on our way over to Duncan Hall to get some chow. See you when we get back."

"Well, okay. Have fun."

Passing out through the gate and straggling along the sidewalk another marching song was taken up:

> *I've been working on the railroad, all the livelong day,*
> *I've been working on the railroad, just to pass the time away . . .*

And so it went as we walked along Yarnfield Road to the far gate of Duncan Hall.

What are you guys up to over here?" George George said. "I though that you were all on standby."

"Oh, yes, we are on standby. But we got hungry. We are just coming over to the mess hall here for a little while to get some chow."

"Oh, yeah, how are you going to do that?"

"Well, there is always a chow line set up for late travelers, and we are late travelers."

"Well, okay. Have fun!"

The hubbub of our entrance into this mess hall somehow attracted the attention of the on duty mess officer.

"Just what the hell is going on here? What are you soldiers doing in here now?"

"Hello, Sir. We're hungry, and we decided to come over here and get some chow. That's why we're here, Sir."

"Well, you are not allowed to be in the mess hall now. I'm going to call the MPs."

"But, Sir, you don't have to do that. We are the MPs. We are on standby duty, and we're hungry. We may be called out when it's late, Sir."

The mess officer no doubt could see that we were just a little bit high. But since we were the MPs, and we were also stuck on the station just like he was, he said, "Okay. Get yourselves something to eat, and then get out of here as quickly as you can. You had all better get back to your barracks where you are supposed to be right now."

So as it turned out, we did have a very fine victory celebration after all. And singularly, the only disturbance by a group of imbibing GIs, which we almost were called on to deal with, was *us*.

Chapter 62

Tales From the Barrack-The Good Old Army Life

Early on, as soon as the 1257th was situated in the Howard Hall barrack, separated from the rest of the station permanent party, there had been an objective to keep the traditional army routine. This intent was mainly achieved by mounting frequent early morning formations of the company formed up in ranks for roll call and inspection.—This was the real army away.—Nevertheless, the number of men present at each of these assemblies was always somewhat less than the entire company for a number of reasons. There were always some MPs who were on duty at the station gates and guardhouse. The ranks were thinned when some men were detached and relocated to Nelson Hall for continuous gate duty. Also, after a time and somewhat contrary to old Army tradition those soldiers who had been on latenight or very early morning guard duty were allowed to sleep in. The duty assignments were always rotating with some men on duty around the clock and many more on nightly patrol duty. Because of all of the around the clock operations there would always be one quarter of the company (one platoon) off duty. So usually there were some that might be absent from the morning formation having not yet returned from an overnight pass. Additionally, in the company there were a few who had relatives in the UK and so early on were authorized to take a furlough to visit these family members. Later, as time dragged on even those who did not have a direct connection to anyone in the British Isles could be granted a furlough. From time to time one or more soldiers in the 1257th were issued special orders putting them on TDY for some specific assignment away from the station. On rare occasions when one of our aircraft was down in the local area one MP was always on guard until it was recovered. This continual reduction of men to muster resulted in the decreasing frequency of early morning formations until they dwindled and finally vanished altogether.

So, when the early-morning assemblies, with their inspections, finally ceased haircuts became less of a priority. And anyway who would know or care if our hair was cut short, because when on duty it was always hidden under our white helmet liners. As with the haircuts we always tried to do things our way and not the Army way. Now, in most cases the Privates and all but the most senior NCOs performed the same duty side-by-side. So generally on our own, usually without direct supervision, we performed our duty our way. Operating as it did, this unit of military police developed a more independent approach to life in the Army than most soldiers could ever dream of getting away with. Off-duty we were a less than finely honed bunch of military police as can the seen in the following photographs. Look at the haircuts, the absence a head covering and the general military bearing. Strange as it may appear these GIs where members of a organization cited as a Meritorious Unit.

MILITARY POLICE OFF DUTY

MILTON COOK

Conversely the same GIs were real soldiers with a sharp military presence when they were on duty. And yet no mater what they did there always was that somewhat independent air about these military policeman. These then were soldiers who performed their duty without totally being caught out in the good old Army life.

MILITARY POLICE READY FOR AND ON DUTY

Chapter 63

Tales From the Guardhouse-Among the Dishonored Dead

In January 1945, there was posted in the Guardhouse a copy of a Special Order from Headquarters, European Theater of Operations, United States Army. This order, signed by General Dwight D. Eisenhower stated that:

PRIVATE EDDIE D. SOLVIK, 36896415
Company G, 109th Infantry, 28th Division
United States Army
Was on 31 Jan to be shot to death with
Musketry for twice deserting before
the enemy.

This would be a harsh event, in a harsh time, which intrigued me for two reasons. First, the U.S. Army had last used muskets as weapons in the American Civil War. Muskets were no longer a part of the U.S. Army's standard ordinance and certainly were not available in Europe in any case. The musket, a smooth bore weapon, lacked both range and accuracy when compared to a rifle. The performance of the musket was so poor that repeated volleys might be required and were SOP to achieve the firing squads objective. In any case, I assumed that the Garand M-1 rifle would be the weapon employed for the execution. The M-1, as it would turn out, for technical reasons, was not a particularly better weapon for use by a firing squad. A basic stipulation for firing squads was that one of the twelve weapons randomly be loaded with a blank. This was to provide assurance to each member of the firing squad that they might not have shot the soldier who was being executed. But the M-1 had a serious kick and with out a muzzle restrictor it probably would not eject the spent cartridge case. The bolt was forced back by the muzzle gas flowing through a small hole near the end of the rifle barrel. However, when there was no bullet in the barrel the pressure would be to low to move the bolt and to generate a "good" kick. So then every one would know who had fired the rifle that was loaded with the blank. The term "shot to death with musketry" has the mystique of an old traditional ceremony, unlike the harsher sounding actual "execution by riflery." Musketry sounds much more poetic than riflery, but the results are the same.

The other interesting thing about the order was that I visualized Private Slovik running away from the front line. Then he was caught, returned to the front line, and he ran away from it a second time. This is somewhat akin to an event, which was related to me by a GI who was a buddy for a short time along the way. He had been an infantryman for a time, and in an action had shot an enemy soldier running away from the German side of the line. In this case, my friend had spared the German Army the bother of a court martial and a firing squad.

Geoffrey Perret in *Eisenhower* states, "there was no evidence to show that Eddie Slovik ever formed an intention to desert." He makes the point that the difference between AWOL and desertion is that the deserter is an AWOL soldier who does not intend to return—*I don't want no more of this Army life*. But, Perret does not recognize that in the case of Slovik, the charge was for the more serious type of desertion, which was punishable by death; to avoid hazardous duty.

In *Eisenhower (A Soldier's Life)*, Cario D'Este also does not clearly differentiate the two categories of desertion from the Army. But he does note that Slovik's "intent to desert was questionable." Both of these authors cite William Bradford Huie's *The Execution of Private Slovik* as their primary source of information for the Slovik case.

Huie was a writer and associate editor of *The American Mercury* and later became its editor and finally its publisher. Huie was the author of a number of books, novels and journalistic treatments of some landmark incidents. His book *The Execution of Private Slovik* has become a collector's item. The used book dealers whom I first contacted all know of the book, but none had a copy of it in their store. I obtained an expensive copy (seventeen dollars, plus shipping and handling) of a 1954 Signet Pocket Book[71] edition of *The Execution of Private Slovik*. This copy of the book is deteriorating and is only a few years short of crumbling. The once white pages have turned to brown; light at the center blending to a darker brown around the edges, which are also prone to splitting.

Huie writes[72] of his visit to special Plot E, across the road from the Oise-Aisne cemetery at Fére-en-tardendis. Plot E is unhallowed ground, the final resting place for ninety-six dishonored dead. Former soldiers, who were dishonorably discharged from the U.S. Army and then expeditiously executed. The ninety-five of these men who were convicted by General Court martial for the crime of murder or rape or both have broken necks, having been hanged. One man, Slovik, convicted by General Court martial of desertion, has eleven bullets in his "heart" having been shot by a firing squad.

The Court martial of Private Slovik was somewhat perfunctory. The elapsed time from convening the Court martial, returning the verdict and adjourning was an hour and forty minutes. The required sentence of execution by firing squad for desertion to avoid hazardous duty was imposed. The Court did not believe that Slovik would actually be shot and that his sentence would be cut to some period of confinement. This was because it was the regular practice to reduce the convicted deserter's death sentence to some period of confinement. During the war campaign in Europe, forty-nine soldiers were court-martialed for desertion and sentenced to death by firing squad. In forty-eight cases, the sentence was reduced to some period of confinement. In Slovik's case, due to the then current encumbrances of war, the "system" went awry and an actual execution by firing squad was approved.

But had Slovik actually deserted two times, much less before the enemy? The fact is that Slovik was never on the line before the enemy. The first time that he "deserted before the enemy" he became separated from a group of replacements on their way to the line. It was night-time; they came under artillery fire and so took cover in

---

[71] Paperback books were a more or less standard size of four by seven inches and were originally called pocket books.

[72] In what follows the specific information relative to Slovik is from Huie's book.

fox-holes. In the confusion, Slovik probably did not hear the order to move out, and so was left behind. Without the second "offence," he most likely would have been classified as a straggler; a soldier behind the line who became separated from his unit for a time. The second time that Slovik "deserted before the enemy" is even more difficult to establish as it was based more on intent than fact. Finally joining his company, which was behind the line at the time, he inquired of his company commander if he could be tried for having been AWOL. The CO had Slovik placed under arrest and returned to his platoon. Later Slovik asked, "If I leave now, will it be desertion?" and later left the area. Once back in custody, in a written confession, he said that he had run away and would do it again. The general consensus was that Slovik, like the other forty-eight deserters, was trying to avoid combat. Also, it has been thought that he had overplayed his hand with the written confession. The actual charges against Slovik were twice violating Article of War 55—Desertion to Avoid Hazardous Duty. Which in his particular case was "combat". Now, hazardous duty is not just "combat". Carrying live artillery shells cradled in your arms comes to my mind, and then just eating in the Mess Hall could sometimes be hazardous duty. So, "desertion before the enemy" was also just another poetic device, really meaning desertion to avoid hazardous duty, which in this particular case was "combat". The term "desertion before the enemy" does tend to influence the perception of just what had occurred, which is why when I read the Special Orders, I did believe that Slovik had run away from the line on two separate occasions. The moral here is, and history is replete with examples, that it is easy to hamper "justice" when fine sounding words are used instead of the harsh ones that actually apply.

When all of the reviews and all of the recommendations had been completed, the final determination was that Private Slovik must be executed by musketry for desertion twice before the enemy. Ironically, with this determination, Slovik achieved immortality of a sort, an honor that more than likely he would just as soon foregone. There is no doubt that Slovik was a reluctant soldier. But, then, most of us were. For most of us, it was a matter of duty and of honor—in the sense of respect, and later of honor—in the sense of pride of unit. Although we never would have said or admitted such, except for pride of unit. In Slovik's case, he had served time in jail as a youth, had been classified 4-F, and then reclassified 1-A when the exigencies of war pressed for more "men," and then had been drafted into the army. Duty and honor were not important ideals in Slovik's background and experience. Training as a replacement infantryman did little to imbue the ideal of duty and honor into Slovik's sense of self. And, of course, there was absolutely no hint of the ideal so necessary to a good soldier—pride of unit.

In the French mountain town of St. Marie Auxmanes, on a bleak midwinter day, the last day of January 1945, they led Slovik along a path dug through the snow to a post erected in the walled garden at number Eighty-six Rue de General Dourgeouis. First bound and strapped to the post, then a black hood was pulled down over his head to complete the preparation of Slovik for his execution.[73] The twelve "picked" soldiers, the firing squad, made their way along another path dug through the snow, halted and turned to face Slovik. These soldiers, his comrades of the 28th Division, with whom and in which he had never really served, obeyed their orders and on the command shot Slovik dead.

Sometime before they had all gone into the garden, Slovik had asked the Chaplain to tell the firing squad to shoot straight and get it over with fast. After the execution had been performed, one of the riflemen said that if Slovik was a coward, he certainly didn't show it today. To which the Chaplain replied that Slovik was the bravest man in the garden this morning. There are many, who when faced with certain and soon death, rise to the occasion. And such was Slovik, who at the end had died with dignity, which we also expect of our heroes.

---

[73] At this point, Slovik was technically a civilian having been dishonorably discharged from the Army of the United States. So was Slovik executed under the Articles of War or under martial law?

CHAPTER 64

Singular Service-Orders to London Town Twice Again

After the war in Europe had ended on 8 May 1945, the Repple Depple continued to process personnel at maximum capacity. Only now, the maximum effort was focused on the repatriation of the GIs, who were now arriving at the station, back to the United States. This Army in the field was departing from the field, and when the last soldier had been processed and sent on his way, the Repple Depple Permanent Party would pack their barracks bags, turn out the lights, and follow closely. At least that is what was believed at that time.

It was good news when the war with Japan ended on 19 August 1945. There might now be somewhat more ocean transportation available with time. But, overall the end of the war in the PTO would not have a significant effect on the operation of the Repple Depple. This was a heady time—at least for all of the GIs streaming in from Air Force units scattered across Europe, who were now on their way home. It was rumored that there was an around-the-clock crap game in progress in the common room of the last barracks up the line. It was a game where monies from across Europe and big stakes were wagered and won and lost. There was a level of excitement throughout the Repple Depple, especially for those who were soon on their way back to the U.S.A. and what was a brand new world. For those of us whose job it was to still man the station, it was just the same old routine. Except now, the nights were bright with automobile headlights beaming, street-lights gleaming, and light from the unmasked windows streaming.

But even with the prospect of to be soon returning back to the good old U.S.A., some soldiers still managed to go AWOL and maybe even miss the boat. Was it just one last fling with some English Darling, a lack of common sense, or maybe they had a reason for not wanting to return home that motivated these GIs to overstay a pass. Whatever their reason was, I had to go down to London to escort some dumb GI back to the Howard Hall Guardhouse two more times after the war was over.

After my last trip under orders, the Repple Depple had been reorganized. Colonel Rader was no longer the CO but was now commanding officer of Headquarters and Headquarters Company 70th Reinforcement Depot (AAF), for whatever was the function of this unit. The Repple Depple proper was now under the command of Colonel John E. Clyde. The new unit designation was as follows: Headquarters; AAF/ET Reinforcement Depot (PROV) (MAN); APO 652. At least the Army Post Office designation had not been changed.

The sets of Special Orders issued for my last two trips were very similar. Both sets consisted of three stapled together black carbon copies that had been typed on eight inch wide by six and a half inch long light-weight gray semi-translucent paper. All three copies of each set had been embossed with the "Official Headquarters AAF Station 594" seal. One thing of interest was that both sets of orders were signed by, the newly appointed, Actg Asst Adv Gen Dorothy A. Kimmel 1st Lt., WAC.

While, the Repple Depple had always been almost entirely staffed by amateurs (draftees), the more experienced non-professional soldiers had learned how to do things the Army way from some of the "old hands," but mostly from the grind of prolonged service. The extended time in the Army of these "old" semi-pros resulted in many of them having an over eighty-point count, which put them in the first group to return to the states. The less experienced of the staff that remained, now had to prepare my out of the ordinary Special Orders on there own without the guidance of the "old hands". So it was not too surprising that anything other than the pro forma dispatching of troops home would cause personnel some problems, even when the "old hands" were still on the job some small errors usually occurred in the preparation of my non-routine Special Orders for Escort duty.

My Special Orders Number 38, Line 11., dated 30 August 1945 had some strange errors. The parenthetical number (67) following my serial number should have been (677). The classification number for "enlisted" military police is 677. The error was most likely a typo on the extract copy of the orders. Next, "47" appears, which is very strange as CMP (Corps of Military Police) should be typed at this point. This was a continuing problem as "CMP" only appear on two of my prior sets of orders. It was just "MP" on one set and AC (Air Corps) on two other sets. But, I guess getting it right two times out of seven ain't bad, considering that military police were a minority of those dealt with by personnel. The significant error in this set of orders is that the number of days TDY allowed to make the round trip was not specified. We (the desk sergeant and I) did assume that the TDY was supposed to be the usual two days.

There was something new now, information that was not in the first five sets of special orders. These two new post war sets of orders now provided the address of where I was to go in London—UK Base, 5 North Adley Street, London. Now that the war was over, there was no longer a concern that German Intelligence would somehow intercept my orders and thus determine the location of the London U.S. Army, UK Base. Another interesting thing in these orders is that "reimbursement for quarters while on TDY in London is authorized only upon certification by billeting officer that government quarters were not available". This was no doubt directed to the commissioned officers; those officers who just might very well decide to stay in a hotel instead of an Army billet. It smacks of fixing an ongoing problem of commissioned officers shacking up in hotels with their English darlings. In my case, this billeting restriction meant absolutely "nothing" to me, as there was always some empty bunk available to billet an "EM" GI for one night.

My seventh and last set of Special Orders Number 58, Line 8., dated 29 September 1945 were almost correct. On this set of orders, I am back in the AC, but the mystery designation has now changed to "568." The work force personnel were shaping up as these orders now allowed two days TDY to make the round trip to London.

Both of theses sets of Special Orders were accompanied with War Department Railway warrants for the train tickets, but there were not any filled in train schedules provided with the copies of the orders. By now, I did know how to figure out the train schedule on my own, so getting to London was not a real problem. That is I knew how to hang around the railroad stations waiting for the trains that I had to catch to get to London. On the return trip, there was Army transportation for my prisoner and myself to Euston Station in time to catch a particular train.

There was really nothing out of the ordinary from what little that I can recall on these last trips to London. Both trips followed the same old routine. Catch a train in Stone and change to a train to London in Stafford. From Euston Station, take the underground to the Bond Street Station and then walk to North Aurey Street. Report in at the Guardhouse and be assigned a billet for the night. Using whatever time was left in the day, do some sightseeing, and then see a movie or a stage show. Then, the next morning take custody of my prisoner and ride to Euston Station in an Army Vehicle. Catch the train to Stafford, and wait in Stafford for the train to Stone. Then, walk down to Stone baggage to call for a ride back to Howard Hall, and turn over custody of my prisoner at the guardhouse. Then, the next day, it was back to regular duty, wherever and whatever time that might be.

Those two GIs, who I had escorted back from London, had by going AWOL most certainly extended their stay in England or maybe in Europe. Chances are that both of those soldiers missed their scheduled trip back home to the United States. So whatever it was that they did while AWOL should have been the basis for a really fond memory, but it probably wasn't.

My engagement with Special Orders escort duty may have been somewhat unusual, but I do not really know because no one ever told me anything except what to do. That I was set off under my own recognizance probably meant that it was expected that I would perform my assignment and that these trips, except maybe the trip to Paris, were looked on as routine menial duty fit for a Private. But, regardless, in each case on each trip, I was just following orders.

CHAPTER 65

Departing From the Barrack-Good Bye Howard Hall

By mid-October, the Repple Depple had processed and sent on their way the last of the repatriating GIs to a port of embarkation for the sea voyage back to the U.S.A. So, now at last, it was time for the Repple Depple to close up shop in England. Camp San Francisco, which was situated at Chateau Thierry, France, was to be the new location of the Reinforcement Depot. This camp, a tent city of winterized canvas shelters with concrete floors and stoves, had been the home of the Army Ground Forces 6069th Repple Depple. Beatty Hall would be closed on 20 October, which was the same day that the Repple Depple Headquarters would officially transfer to Camp San Francisco. On the first of November, Howard Hall was to be vacated. The last, on the 15th of November, the Army Air Force would turn off the lights and depart Duncan Hall. Thus would end the United States Army Air Force occupation in Britain during World War II. There would still be units of the Army Air Force (the soon to be United States Air Force) stationed in England, but the Army in the field was departing. Coincident with the departure of the Army Air Force from England, the Army Service of Supply was also shrinking and departing from the field. The dwindling of the SOS resulted in some logistical problems for both the delivery and the amount of food delivered to the Repple Depple. This meant short rations for the last part of our stay in England. There were, some left over K-rations available which provided an interesting supplement to the somewhat meager chow.

As the rest of the Repple Depple staff shrank so too did the 1257th dwindle away, as those with the longer service time, and thus a higher point count, departed for home. Soon only about fifty of the sixty or so remaining members of the company were from the core group, which had been in basic training together back at Miami Beach when the 1257th had been formed. The point control post at the junction of Yarnfield Road to Highway A34 had been vacated for some time. Foot patrols in our three venues had ceased. First, patrol duty in Millmeece had ended when Nelson Hall went dark. Then, when the last pass was issued to the last casual GI passing through the Repple Depple, the foot patrols in Eccleshal and Stone came to a halt. On my last pass to Hanley, I had said goodbye to Lillian. We exchanged our addresses, not really intending to write to each other. Exchanging addresses is a social gesture that occurs when good friends part, but who never ever expect to see or hear from each other again.

It was in the last week of October when the Order was issued: Pack up you barracks bags, we are moving out of Howard Hall. This was to be somewhat like leaving home as most of us had been settled into these rooms for two years. Howard Hall was to be closed on the first of November, so we would spend our final days at the Repple Depple billeted in the almost empty Duncan Hall. Coincident with the move to our temporary quarters the 1257th had effectively ceased to exist; we were now members of the 156th Reinforcement Company in the 130th Reinforcement Battalion. The employment of reinforcement units provided for the orderly movement of single and groups of soldiers instead of specific units. This transfer to a "new" unit was not a concern as we would very soon be on our way back to the United States.

But, then in the second week of November we were informed that most of us, instead of being on our way home, would in fact be going to a new assignment. This was a move that would in essence reconstitute the now reduced 1257th Military Police Company. So now we were leaving, but as usual we had no idea of just where it was that we were headed. The next morning after chow, we loaded our bags into one two and a half-ton Army truck, and then we climbed up in the back of another. It was goodbye to the Repple Depple forever as the trucks drove off down Yarnfield Road on the way to Stone. Stone baggage was long gone, and so as it would be for the rest of our trek, we had to take are of our own bags. When the train came to a stop at Stone Station, we stacked our bags in the end of an empty car, which had been booked for our use. Then, we went along the side passage to find a seat in one of the compartments for this final trip south to London.

The barracks bags were loaded directly from the train onto baggage carts, which were rolled through Euston Station and on out of the entrance. Here, we transferred our bags from the carts to the back of an Army truck, and then climbed up into the rear of another one for the trip to Adley Street. Once offloaded from the trucks, we were directed into a room that had been one of the London Headquarters "Enlisted men" barracks. It was a long room with walls of raw soft red bricks set in gray mortar. This large room, which may have at one time been used as a warehouse, was filled end to end with two rows of closely spaced double bunks, each holding a thin bare mattress. The heads of the bunk were set close to the side-walls to create a somewhat wider central isle. This room must have been a real zoo when it was filled with GIs before the invasion of France. Each one of us picked out a lower bunk located near to the entrance of this room. Then we unpacked enough to make up our bunks for sleeping and to get comfortable for the rest of the day and night.

Located in another large room in these London Army Headquarters, the Mess Hall was mostly empty when we were taken there for chow. After the evening chow, there was not much to do except talk, play cards or read a book until lights out. The next morning after cleaning up and visiting the Mess Hall, we repacked our barracks bags and hauled them outside of the building entrance. Repeating the routine of the day before, the bags went into the back of one truck, and we went into the back of the other.

This morning, the trip was short to the railroad station where we unloaded our bags onto the baggage carts and then unloaded the carts onto the train. The train trip this morning was a lot shorter than yesterday's trip, and soon we arrived in Dover. One there, it was transfer our bags and get on board the cross channel ferry. Finally, the whistle blew, and the ferry got underway, heading across the English Channel to Calais. This was a cruise that I did not look forward too as I was, at that time, very prone to motion sickness. But this was only going to be a short journey, and I would be back on dry land in a few hours. So, on this day, we would stay no more on England's shore, but were off on our transit of the capricious channel.

PART
IV
ARMY OF OCCUPATION

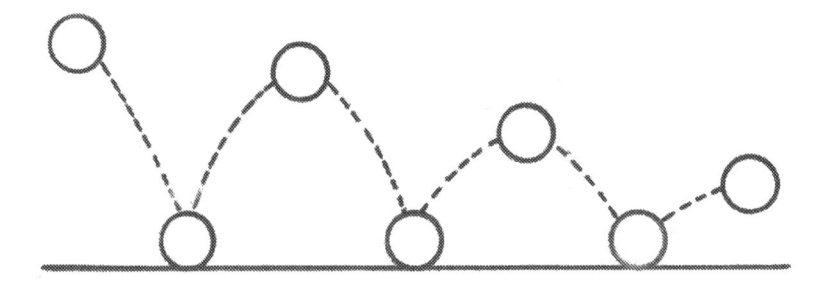

Chapter 66

Off to Germany

Swirling and squawking, the cacophonous colony of gulls greeted the cross channel ferry as it backed down and slowly nudged up against the pier. We were now docked in Calais, France, and this was the first time for most of the remaining members of the 1257th to be on the far shore. After a short time of standing by, the company was instructed to pick up our bags, proceed off of the ferry and go into the terminal building. Once inside, a perfunctory clearing of customs was exacted, after which we were directed through a side door. Once outside, we were told to load the barracks bags into one of the U.S. Army trucks sitting in the parking lot, and then get into the back of another. These trucks had been sent here to provide transport to the next destination on this mysterious journey into Europe. Now traveling along on the right side of the streets and highways, we watched farm, villages, and cities diminish and disappear behind the back of the truck.

After a few hours of this hindsight seeing, the truck finely arrived in Paris and the *Bois de Boloene.* Entering through the gate at the Chateau Rothshield, the truck parked, and we were told to get out now, but not to unload the barracks bags. To our disappointment, we were advised that the chateau was not our final destination and the company had only been taken here for our "noon day" meal. After lunch, there was little to do other than wait until at last the order was issued to get back up into our truck. A very short trip back through the streets of Paris brought the truck to the front of a railroad station, where again we climbed down from the truck and unloaded the barracks bags from the baggage truck. Once again, baggage carts were acquired, as this bunch of soldiers would not haul barracks bags around on their backs when it could be avoided. We were after all by this time "seasoned" troops, who knew quite well just how to beat the Army at its own game. Dragging the now loaded baggage carts, we were now headed towards the station platform to board the train for the next part of this long journey to wherever.

Stacking the barracks bags up in the back of the vestibule, we continued along the side passageway of the coach to pick a seat in one of the compartments. This was a third class passenger coach that would hold comfortably no more than six passengers in a compartment. The two facing seats were very hard, wooden slat benches (with narrow gaps between the slats) that were contoured to, more or less, fit the shape of a human body. On these wooden benches, we would sit and "sleep" as the long day's journey continued through the very long night. In between the two facing benches, a narrow table was connected to the outer wall of the compartment and supported at the other end by a single leg. This small tabletop was a surface upon which to play cards and later a place to eat our evening chow of K-rations, which had been left over from the Second World War. One of my compartment mates had, somewhere along the way after arriving on the far shore, picked up a German Army field stove. This small field stove was an ingenious device made from pieces of thin sheet metal that were hinged and pinned so that it would fold up into a compact unit. With the stove, there was a packet of fuel, small rectangular tablets of a white granular substance, somewhat resembling a cube of sugar. The small brick of fuel was placed in the center of the stove's middle shelf. When lit with a match, the tablet slowly burned as it heated water in a canteen cup. We were heating the water to make coffee using the instant that came in the boxes of K-rations. This solid fuel must have been something like a solid version of Sterno.

Late in the afternoon, the train slowed down, stopped and then sat for a long time. The train was parked at the French/German border, where customs must be cleared and the railroad engines and crews must be exchanged, before it could be on its way again. By the time that the train was rolling again, it was now dusk, and the sun would soon disappear. Shortly after the train was on its way through Germany, the expansive flat fields stretching away from the train were pockmarked with many small craters. One of the GIs in our compartment said that, the craters were the result of exploding German eighty-eight (880-mm) shells fired from tank cannons. How he knew this I cannot recall, but he may have been "there" before he had joined us in the military police. Now it was totally dark. Very few lights were to be seen any place across the countryside as the train rumbled through the night traveling deeper and deeper into Germany. Soon the compartment lights were turned off, and so we sat in the dark and thought about sleeping. Resting your head on your arms on the surface of the table was really no better than trying to sleep more or less sitting upright, and in either case, my butt was hurting enough that it hampered any real sleep.

Now it happened to be that one of the soldiers in the 1257th was prematurely gray, and the German train crew took a notion that he was in fact an old man who somehow had ended up in the American Army. This was not a completely strange idea, because as the war closed in on Germany, both the very young and the very old had sometimes been pressed into service in their army. Probably to honor this "old man," some of the crew took him on a tour of the train that ended up in the cab of the engine. There he was allowed to hold the throttle ad operate the train, and, to some of the train crew's consternation, he ran it much too fast along a not too stable stretch of track which rested on a damaged embankment. We did not find out about this flirtation with disaster in the dark until some time later.

It was sometime between zero four hundred and zero five hundred hours when the train squealed to a stop. The lights in the compartment came on, and one of our sergeants poked his head in to "wake" us up, and told us to put on our field jackets, grab our mess kits and get off of the train. They were turning us out in the middle of the night to feed us breakfast. We blearily scrambled down to the surface of an open strip of coarse gravel, bounded on one side by the train and on the other side by multiple sets of parallel railroad tracks fading into the dark of night. It was roaring cold, and the strip of ground where we were assembling was an oasis of light in a pitch-black desert of dark. High on poles, large enameled reflectors directed the dazzling light from oversized, clear, incandescent lamps to light this space between the runs of railroad tracks, where a U.S. Army field kitchen was in operation. In the heart of this railroad yard was probably the best spot where the trains, both coming and going, could stop to feed the moving Army. For breakfast, it was the typical Army morning chow of eggs of some kind, meat of some sort, canned fruit of some type, pancakes with syrup, and coffee. With our mess kits filled, it was belly up to the high long narrow tables and eat this mornings' chow standing up. When breakfast was finished, it was past the usual triad of GI cans—garbage, hot soapy water and hot clean water—cleaning our mess kits on the way back onto the train. Back in the compartment, we nestled in for what would be the last and somewhat shorter leg of this train trip through the night.

The sun had risen, and it was light when the train ground to a stop in the Munich (AKA München) Railroad Station. This was the end of the line for the train portion of our journey into Germany. Grabbing barracks bags from the pile in the end compartment, we stumbled, bleary-eyed from the train and loaded the bags into the back on one of the waiting U.S. Army trucks. Scrambling up into the rear of another Army truck, we were shortly on our war through downtown Munich. The view from the rear of the truck as we passed through the city was somewhat bizarre. The streets and the sidewalks were clean, neat and in very good condition. But, then just at the inside edge of the sidewalks, there were tall walls—tall walls of neatly stacked loose bricks and rectangular cut stones piled higher than the top of the truck bed cover. Above and inside of these carefully constructed walls, there were heaping mounds of rubble, surmounted by jungles of blackened, bent and twisted steel girders. The Germans had so very neatly cleaned up the mess that had been made night after night by high—flying British bombers.

Once out of the city, the highway ran through forest and farms. After a few hours of rolling along over hills and past trees and brown fields, the truck abruptly turned off from the highway and made a brief stop. Once moving again, the tableau that appeared behind the truck confirmed that the truck had just entered some military establishment. We had arrived at the Fürstenfeldbruck Air Base, the very place where the Luftwaffe had trained its pilots. The 1257th had been part of the U.S. Air Force for two and a half years. Now at last, what remained of the company was to be stationed at an airfield. The truck proceeded slowly past a very long multi-winged two-story building and stopped when it came to its end. The truck driver then got out and went off to find out just where he was to unload us.

This very long structure that we had slowly ridden past was in fact the barracks building. There was a long hallway that ran from one end of the barracks to the other end on both of the floors and the basement. Three level barrack wings ran out from the extended corridor, but instead of putting the company in one of these wings, the 1257th was at first billeted in one large room located on the second floor. This room had been set up with rows of cots, several tables and some chairs. A large latrine was located conveniently on the other side of the hallway. This wide room that we were now all sharing might have been a common room, a lounge, but more likely, it had been used as a staging area for the assembly of mess formations as it was beside a wing on the other side of the central hallway that contained the Mess Hall. This building had been constructed so that all of the necessary facilities would be readily accessible during even the worst of the Bavarian winter blizzards. We were hopeful, because we did not have any winter gear, that our replacements—pardon me, reinforcements—would be here to relieve us before the cold of winter and the snows came. We were, for a short time at least, now part of the Army of Occupation and were stuck here until someone came to take our places, regardless of our point count.

So here we were, but just why had we been transferred to Germany instead of being on our way home at this point in time? It is only conjecture, but probably the dwindling 1257th was the only aviation military police unit still available in Europe. With the exception of the permanent party MPs at the one remaining Air Force Station in England at Burtonwood, we the last to leave and so were all that remained. All of the other MPs had returned back to the United States with their units as each Air Force Station closed. Also, the Army Air Force had conducted little in depth planning for the long-term operation of the long-term occupation of Germany. Concentrating on defeating the enemy little planning had been devoted to just how to conduct the occupation after achieving the victory. (You mean now we have to hang a around here for years and years just to complete the job?) At any rate a decision was made to utilize the Fürstenfeldbruck airbase for Army Air Force operations. (Which as it would turn out to be for a very long time, with the coming of the Cold War.) And so there the was1257th with its job in England complete, but still in Europe and thus available for additional service. That is the most likely reason of how we came to spend almost two months in Germany along the way on our long journey home. As it turned out even though the 1257th was stationed in Germany after the war was over we did not serve in the Army of Occupation. This was a technicality, as a minimum of three months service in Germany after the war was over was required to qualify for the Army of Occupation Medal.

Chapter 67

A Jaunt Through the Town

It was not long after the 1257th was situated at the Fürstenfeldbruck Air Base when Foiles, Szucs and I decided that it was time for us to make the sightseeing tour of the town itself. On this afternoon, we all three were off duty and so taking a look at the town would kill a little time and assuage our curiosity. Changing from our fatigues into Class A OD uniforms, we set out for the guardhouse, to look it over first and then to check out of the base. A pass to visit the town was not required, as there were no passes being issued at this time. In the long term the restriction of the American soldiers to military installations would be an untenable situation. But at this time that mattered little because the German economy was still in shambles. So this was my first and only visit to Fürstenfeldbruck and of the guardhouse for that matter.

My only duty post was at an internal side gate far from the guardhouse. There was not a requirement to go there and check out before going on guard duty, and so I had never been to the guardhouse. But before striking out on our tour of the town, we had a guided tour of this very German guardhouse. This Luftwaffe guardhouse had a number of single prisoner cells. It was solitary confinement with a thermal touch, each cell had its own externally controlled thermostat that was used to vary the internal temperature from well below freezing to above 120° F. I suppose, this was one of the methods used to maintain the strong discipline necessary in a totalitarian military organization.

It was, in a way, very strange walking into a town which only five months prior had been deep in the enemy's homeland. The natives were not at all friendly which should be expected. While they probably did not recognize the USSTAF patch that we wore on our left shoulder, they were more than familiar with what the Eighth Air Force patch on our right shoulder represented. This was no doubt the source of the many hard looks that came our way, because the populous, with significant reason, did not hold the Eighth Air Force in high regard.

The road as it entered the town crossed over a bridge, then turned right and ran along what must have usually been a shallow stream at the bottom of a wide sloping-sided channel. On this day, the trough was filled with rapidly flowing, cold, crystal clear water. On the sloping sides of the ditch, the long green grass growing on the sides was bent far over and waving in the fast flowing stream. Walking along, we suddenly saw at the very bottom of this deep, chilled, clear stream a shining saber. It was very enticing as this blade would have made a very fine souvenir to take home to the U.S., but there was no way to retrieve it from the bottom of the frigid stream. How and why this sword had ended up on the bottom of the channel was a source of some speculation. Another object of interest of this short jaunt was the impressive medieval cathedral, which was so large that it appeared to be out of place in this small town.

Other than sightseeing, there was really nothing to do on this trip to town, because the U.S. Army's policy at this particular time was to limit contact with the German population. A very shortsighted policy, as this would be a long hard occupation. In any case, there was a General Order in force at the time that prohibited fraternization with the German women. I believe that fraternization meant, for the most part social activity; as there had been many reports in the *Stars and Stripes* of GIs being fined sixty-four dollars upon being convicted of the crime of fraternization. The amount of the fine was most probably waggishly based on the CBS radio quiz show, *Take It or Leave It*, which had debuted in 1940. On this show the top money, the big money, and it was at the time, was for answering the sixty-four dollar question.

Chapter 68

Life in the Barracks

The long narrow mess hall ran along side the barracks building in line with the wide wing where the 1257th had been initially billeted. The passageway into the mess hall was in line with the bottom of the stairway that ran down from the side of our wing. The mess line started to the left of the door and then ran along the wall and then turned right to be served from a counter running across the room. Sitting at the mess tables the pine forest, in which the Airbase was located, could be seen through the tall windows facing away from the barracks. Separated by a narrow section of wall, the panels of small square pane lattice windows stretched from waist high to about the same distance from the two-story high ceiling. A short distance below the ceiling, all four walls had been embellished with a continuous stenciled band of Nazi-themed images, such as swastikas and fledgling birds.

There had to be a few U.S. Army cooks in charge of the kitchen; however, most if not all of the actual work in the mess hall, was performed by German POWs and some civilian workers from Fürstenfeldbruck. All German Government and to some extent social functions had been subject to strong central control. It was an ingrained part of the German culture to be subservient to authority. So with the defeat of Germany and the removal of most of the nation's top leaders, all social institutions were significantly disrupted, and food was in very short supply. With this situation, working in a U.S. Army mess hall was a very good deal for both the POWs and the civilian workers, as they would eat well even if it was not their usual fare. But the American Army's bountiful supply of foodstuffs, coupled with the shortage of food in the countryside was a source of temptations. One of the obligations of being a prisoner of war is to plan, work and if the opportunity arises, escape from your captors. A many times tried, and sometimes true method of making a getaway was to dig a tunnel under the stockade fence. Therefore, it was not a complete surprise when one of the MPs checking the POW enclosure discovered the entrance to a tunnel hidden under an anomalous oil drum that just happened to be sitting over it. This tunnel ran from the compound where the German POWs would "escape" and go into Fürstenfeldbruck to visit friends and maybe relatives to convey to them some of the American Army provisions. Then, in a somewhat unorthodox action, completely opposed to the POW code, they would crawl back into quarters before dawn's early light. At this point in time, there was no real advantage in escaping, but even so, the POWs nightly outings into town were terminated forthwith.

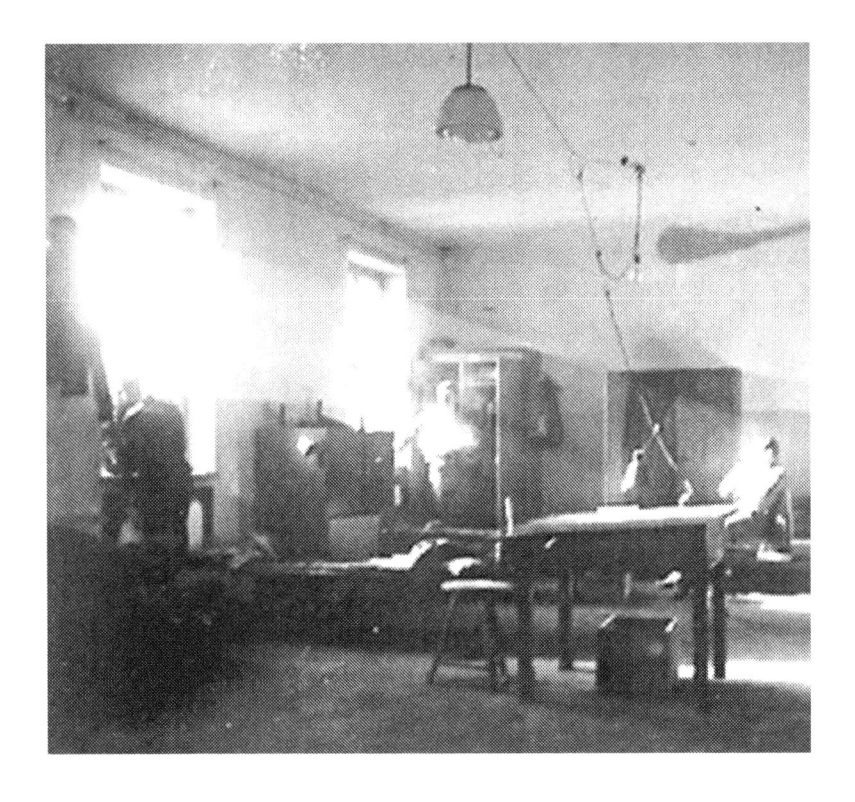

FIRST BILLET IN LARGE ROOM

THE STAIRCASE THE HALLWAY

In another instance, one of the MPs on regular gate guard duty became suspicious of the shape of things. His observations left him to suspect that one of the German civilians was sneaking out an egg a day, nestled in her bosom when she left the Base. One day, to break her of her smuggling ruse, he gently snapped the outer edge of his hand between her breasts. He told me that neither one of them said anything, but that there was a very scrambled expression on her face as she walked on through the gate.

Always exploring, one day I discovered that the basement in one of the barrack wings was almost completely filled with a jumble of German ordinance maps. I did think about unrolling some of theses maps to see if I could recognize the country and place charted. But then I figured that it was just too much trouble as the chances were small that I would be able to identify the country from the small area mapped. On another day, when nosing around, I found a small rectangular piece of mirror glass, which more than likely had been part of an aircraft gunsight. I could see my reflection and at the same time see what was behind the mirror. This small pane of coated glass was very interesting and was a nice compact souvenir to take home. Then one day, The Bad Penny borrowed my "mirror" to use for shaving. But why he did this, I'm not quite sure, because there were mirrors in the latrine. In any case, the mirror was broken when he dropped it, and that was the end of my souvenir.

At this point in the occupation of Germany, most of the barracks in this building were still empty. The few GIs billeted on this part of the Air Base were all clustered near the mess hall. So, for convenience, one of the close by first-floor barrack had been converted to the PX and the PX store room. In another close ground-floor barrack, a Red Cross canteen had been set up. Now, in England, except for management and equipment, all staff and supplies, whenever possible, were obtained locally. But in the case of this Red Cross Canteen, the reading material and the donut ingredients had to be shipped in by the Army. One exception was the "coffee" which was obtained locally. This was the German ersatz version, coffee in name only, which was concocted from roasted barley and chicory. It tasted like coffee, and it was as good as some of the Army brews that I had drank on occasion. But this was a different sort of coffee, because whenever I drank any, it never lingered long and soon would have to be passing on its way.

Chapter 69

GI Cough Drops

When I was a boy and still as a young man, I frequently had a bronchial complaint. This condition most probably stemmed from the fact, as it was related so many times, that as a baby, I had almost died of the croup. In any case, in my younger years, I frequently had recurring bouts of extreme irritation of the throat, which was always accompanied with a racking cough. And now, it was all back, a loud racking and frequent cough, which was way beyond annoyance, disturbed the sleep of everyone (most of all, me) who bunked down in the common barrack room. Early on, after what was a really bad night, the universal consensus was that I must go on sick call this very morning.

Therefore, my mission was clear, so I set out for the infirmary to report on sick call. Off I went, down the stairs to the barracks-spanning corridor and out of the north end doors. Proceeding for only a short distance, I came to a halt at the street curbing. I had to stop at this point for the very reason that a bus was passing by along the street. This bus surprised me, as I had no idea at this point that this Air Base was so large that it had its own internal transportation system. Standing on the curbing, I watched as this bus proceeded ever so slowly past me and went on its way. Besides traveling so very slowly, the other strange thing about this bus was the small trailer that it had in tow. It was a very unusual sight. The trailer had on it what appeared to be a stove with a chimney taller than the bus, a pile of firewood and some other strange operating apparatus. Among the several lines running between the trailer and the bus, one appeared to be a hose. From this, I surmised that the equipment on the trailer was some sort of a methane cooker that was producing the fuel on the go—fuel to keep the engine running as the bus proceeded ever so slowly on its way along the street.

Crossing the street, once the bus had passed, the hospital was visible at the center of a large flat field covered by a short faded green grass. Except for the few scattered around the infirmary, all of the other trees were precisely trimmed and spaced in a line running around the perimeter of the field. I thought that this large flat lawn just might have been used as a drill field. I imagined the fledging Nazi airmen goose-stepping back and forth across this parade ground as they practiced close order drill German style.

Reporting for sick call, my coughing quickly caught the attention of, and established a priority with, the on-duty medic. His diagnosis was straightforward. He quickly provided me with a small packet of troches and the instructions to dissolve one in my mouth whenever I started coughing. These dark brown cough drops were about a half-inch round by a quarter inch thick with slightly convex ends. These lozenges tasted very, very bad. As the saying goes, they tasted like the other brown stuff that they somewhat resembled. However, these troches did have a somewhat curious side effect in that they quickly and completely stopped my coughing, so I was spared from having to dissolve very many of these GI cough drops in my mouth.

Chapter 70

Boys Will Be Boys

Recreation was "do it yourself" as the Air Base was not yet completely operational, and a trip to town was out of the question. The more than usual boredom and tedium was blunted somewhat by the usual activities of playing cards, reading, listening to Armed Forces radio and an occasional visit to the Red Cross Canteen. Still with only four hours of duty every day, there was a lot of time to kill. So we might spend some of the free time exploring the close by areas of the Air Base and that part of the countryside that was within its perimeter. Close to the barracks, there was a large two-story administration building that was locked up because it was not yet in use. So after walking all around it and under it on the roadway that ran under one wing, it was of little interest to us. Away from all of the buildings and into the countryside, the area was park-like with a small lake and a somewhat decorative garden. The only evidence of possible combat on the Base that we ever found was one partially demolished small building. I was not clear if this building had been hit with some type of ordinance, so we were left to speculate on what had caused the damage.

Then one day, as we wandered in the woods, Szucs and I made an intriguing discovery. It was an open gravel field strewn with the detritus of war. A spot where just a half of a year before, around the first of May, the tide of battle had ebbed as the war came to its end. Scattered about in this clearing, we found many and various parts of U.S. Army uniforms and ordinance. Looking over the clothing, I discovered a pair of perfectly good GI coveralls that fit me. After washing them, I would wear the coveralls sometimes just to be a little different, a non-uniform uniform. But my real find was an OD knit baseball cap. The cap was only a little tattered, and it did fit my head,

THAT'S THE WAY THE BALL BOUNCES

HEAD QUARTERS BUILDING AND RELIEF

AT LEASURE ON THE GROUNDS

so I would wear it most of the time when off duty. This cap was lost at sea when it fell from my head to a lower, off limits deck of the troop ship on which we were sailing back to the United States. But the ordinance was much more interesting from the point of view of our experience of never having seen these items before during our total tenure in the Army Air Force. We discovered a Vairy pistol that was used to shoot various colors of signal flares up into the air. Several short sections of fifty-caliber linked belt machine gun cartridges were also lying there in the field. These were very interesting artifacts, and we took each other's picture as we played around with the ordinance. I kept one fifty-caliber cartridge as a souvenir, from which I subsequently removed the gunpowder. Based on the type of materiel strewn about on the gravel, I surmised that some U.S. Army armored unit had last occupied this field.

COOK AND SZUCS IN ACTION

On another occasion, Collins said to Foiles and me, "I've got an ammo can full of loose forty-five cartridges. So, get your automatics, and we'll go out to the firing range and do some target practice." As usual, not really having anything of real interest to do, we both replied, "Okay, let's go." We slung on our side arms and took off with Collins leading the way, as he was the only one of us who knew the location of the firing range. The three of us had been inducted into the Army in Los Angeles on March 8, 1943, and had served together ever since. As we walked on the way to the firing range, Collins told us that he had enlisted in the United States Army and would be staying in Germany. Foiles and I were both very surprised at this disclosure and thought that it was sort of a dumb thing to do, because, for ourselves, we could hardly wait to get back to the States and out of the Army. Now, Foiles had a wife and a life to return to, but those of us who were somewhat younger did not have any kind of an established life to which we would soon return. It may well have been that this was Collins's situation, and he saw the Army as a good opportunity at the time. In any case, he must have received a good promotion in his rank for enlisting and staying on in the Army of Occupation.

The firing range was dominated by a substantial concrete backstop. Three short wings projected away from this tall thick wide wall. The upper ends of the two oblique sidewalls were connected to a concrete awning that curved back towards the firing range. These three wings had obviously been shaped to deflect any stray or ricocheting bullets back into the firing range. A shallow, thick concrete shelf, somewhat resembling a stage, projected from the backstop and wings at about head level. This ledge was just the right place to set up our makeshift targets. This shelf was probably much higher than it had been during most of the backstop's existence. The earth sloped up to become a mound a short distance in front of the deep backstop footing. I may well have been that during the latter part of the war, this ground was dug up in an effort to recover and recycle the spent slugs from firing practice.

Once our targets were set on the ledge, we commenced firing away as fast as possible, stopping only long enough to reload clips and then firing away again. We tried shooting from hip, cowboy style, but the automatic, we found out, is not really suited for that particular usage. It was just somewhat disconcerting having the spent brass and the gas flashing past my face when the gun ejected. We had been trained to aim and then carefully squeeze off the shot when we shot our automatics. But, I found that for normal short ranges, aiming high and to the left and then jerking the trigger worked just at well. I had spent countless hours, when there was little else to do, exercising the trigger mechanism of my sidearm to wear it in and obtain smooth action. This ritual was just in case I was ever in a situation where I would actually be required to fire my sidearm. But, strangely, after basic training, we never had firing practice with the .45 automatic. The 1257th had on several occasions been taken by truck to an unused rock quarry where we fired at boulders to maintain our "expertise" with the M1 Garand rifle. But never did we practice, or even fire, our primary weapon, our .45 caliber sidearm. Much too late, after the war was over, I read in the Repple Depple newspaper that there was a pistol range where some of the permanent party had qualified on the automatic. It is a mystery why the MPs were never qualified on their primary weapon, the .45 automatic.

After a while, we were completely bored with all of the target practice and just quit even though there was still a lot of ammo in the can. With all of the shooting which we had been doing, it seemed somewhat strange that no one ever came to find out just what was happening. Maybe Collins had informed the guardhouse that we would be dong some firing out on the range. Then, maybe the firing range was in a location where the sound of firing was suppressed. In any case, this was the only shoot out with my sidearm that I ever experienced during all of my service in the United States Army Air Force.

Chapter 71

Guarding the Gate 'Till the Very End

It was a low and very sturdy concrete block wall, maybe a meter and a half tall by ten centimeters thick. Demarcating the separate area of the Air Base the wall was not high enough to provide any kind of security, and besides the two-lane opening in it was not closed off with a gate. This wall more than likely ran all the way to the roadside perimeter of the base, but I never followed it far enough to find out. Past the gap, the wall continued on for a ways before it turned right and headed off into the woods. This corner was where I would first encounter the wall every day as I walked along the road running past the barracks to my post in the guard shack located just past the "gateway."

Probably leaving some space for parking, rows of long building were set well back from the other side of the wall. The two groups of building, which were separated by the street, ran parallel to the wall. The long, narrow, single-story, peaked-roof buildings appeared to be either warehouses or maintenance and repair shops. However, at this time, there was no activity that could provide even a clue as to just what had been the function of these buildings. In any case, to control access between these buildings and the rest of the Air Base, a brand new gatehouse had been built. This guard shack was next to the wall on the side of the gap towards the front of the Base. It was a small hut with only enough room for two MPs to sit comfortably on stools. From waist height to the roof, the four walls and the door were mostly glass. With grand foresight, an electric heater had been installed in the shack to assuage the cold of night and the coming bitter cold of winter.

This guard shack would be my post where four hours a day, I would watch the gap in the wall. But nobody or nothing ever passed through the gateway during the whole time I spent there on duty. It may well have been that when activity eventually picked up on the Air Base, this guard pose would in some way serve a function. But, during the time that I was on duty there, it appeared to be a "make-work" guard post.

When I arrived for my very first time on duty at the guard shack, it was not quite completed. Entering the hut, I saw that an old German carpenter, a craftsman, was still at work. He was installing molding around all of the windows. I watched as he hammered in and then counter sunk each finishing nail. Next, he filled each small hole and carefully wiped off the excess putty before painting the molding. This attention to detail where it is not really necessary was an intrinsic facet of the German culture. This attention to detail probably consumed countless hours and contributed little to the German war effort, which was of course in our favor. This is in contrast to the American system of just good enough[74], where molding would most likely not have been installed and certainly not with counter sunk finishing nails. As this carpenter worked away we tried to converse with each other. With his Bavarian German and my American English, we could almost communicate owing to the residual Saxon lineage of our disparate languages.

Early in December, it turned very cold and started to snow. Luckily, it stopped snowing after only a few inches had fallen. So once again, it was time to implement the innovative way in which I dressed for the cold. First, two pair of socks, then a pair of cotton fatigue trousers inside a pair of OD wool trousers with the bottoms of both secured to the tops of my combat boots. Next, my OD wool sweater vest over an OD wool shirt, then the gray RAF scarf around my neck before putting on a field jacket. Now, I had an OD wool baseball cap to wear on my head under the white helmet liner and last, a pair of OD wool gloves. I was warm walking along the road to the guard shack even though it was so cold that the leather soles of combat boots squeaked with every step on the snow.

Arriving at the gatehouse only a few days later, I was surprised to have a new duty partner, who was there for on-the-job training. He was a young GI, a draftee fresh from training and from the United

---

[74] Sometimes stated as "good enough for government work."

States, who had been too young to serve in the war. But, now he was old enough to be here in Germany and be part of the Army of Occupation. It was very timely that at last our replacements were arriving neck and neck with winter. So, hopefully, we would be on our way home before the weather got really bad, and the snow got really deep. Several days later, upon returning to the barrack, from what turned out to be the very last time on duty in the Army of the United States, I was told to start packing as a group of us were shipping out the next morning.

PART
V
HEADED HOME

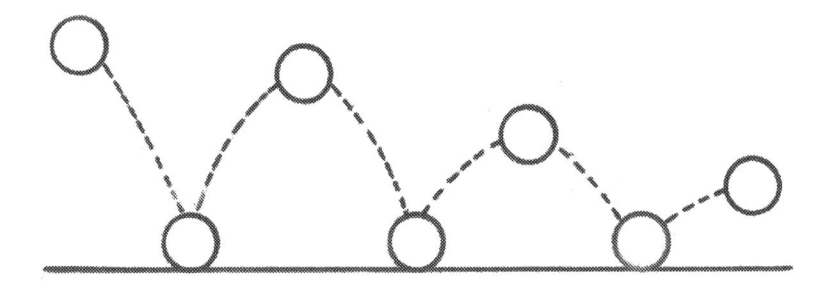

Chapter 72

The Last Time I Saw Paris

Soon after I returned from our morning chow, instructions were given to finish packing the last of our gear. Then, when packed, we were to carry the bags out of the barracks and load them and ourselves into the back of the truck parked by the north entrance. After loading up, we sat in the cold until finally the truck driver showed up. Then we were quickly on the way for a cold trip to Munich and the railroad station. At least the coach was warm after we eventually boarded the train. Our comfort was tempered by those third class wooden benches, which were still just as hard on the butt. After an almost endless ride, the day and the journey were finished as the train slowed to a stop at the end of the line in the Paris railroad station. Once outside of the station, we found the U.S. Army truck that had been dispatched to meet our train. After loading onto the truck, it was only a short trip to the *Bois de Bologne* and the Chateau Rothshield for a late supper.

When it had passed quickly through Paris, almost two months before, the remnant of the 1257th had stopped in the Chateau for lunch. This time an even smaller band would be staying on for just a little while longer. It's really uncertain if our journey back to the States was planned to allow us a few days in Paris, or that it was just the way that it all worked out. I think that our journey had been scheduled so that we would spend a few days here as a kind of reward to compensate for the diversion of the 1257th to Germany. More than likely, there had been many GIs in Europe who never had a chance to visit Paris. But in any case, I was happy to spend a short time in Paris for whatever the reason.

Earlier, near the end of March, Foiles and I had flown from England for five days TDY in Paris. So, now back in Paris, I did have some idea of how to get around the city. However, as it turned out, this visit to Paris much surpassed the first one, as I fell in with a good companion who knew the city inside and out. Herm Martell had been stationed in Paris sometime before. I think that his assignment had been some form of liaison duty. Martell and I hit it off and were immediately best buddies. Soon we would both be headed for opposite sides of the country (he was from Brooklyn), so it would be a short friendship at best. We did exchange addresses and later letters, but as time would pass, the connection that we had for this short time soon faded away.

AT EASE
Chateau Rothschild
December 1945
In the latrine
In the garden
But not
Flagrante Delicto

Martell told me that when he had been stationed in Paris, he would ride the Metro from his billet to his duty post. Every morning, before he boarded the Metro, he would buy a newspaper to read on the way. But, he never was able to finish reading the paper because as soon as he started reading, some Parisian would ask if he spoke French. It was the same every morning with some fellow passenger engaging him in conversation until the Metro stopped at his station. Being fluent in French as he was is why I thought that Martell must have had some liaison function when he was stationed in Paris.

Ranging over Paris with Martell was a singular experience as we went places and saw things that I would never have contemplated doing on my own. With Martell speaking French like a native, we had the best service wherever we went, be it bistros, cafes or the hot nightspots. I remember well one night, as we strolled along the boulevard and encountered a street vendor roasting chestnuts over a charcoal blazer. Now if I had been with anyone else, we most likely would not have stopped and bought any chestnuts. I had never eaten any chestnuts before, much less hot fresh roasted ones, but they were a pretty darned good treat.

Then on one evening as we wandered from one place to the next, Martel told me a permissible story. At present, he was in the Army Air Force, but originally he had been in the Infantry. His company had been on the line for five or six days when he was hit. After the medics attended to his wound, the stretcher-bearers carried him back to the aid station. The nurse looked down at him and asked, "Soldier, why are you clean shaven?"

"But Ma'am, I'm only eighteen."

The nurse cried.

Chapter 73

I Won't Be Home For Christmas

The Christmases in England in '43 and '44 was much the same as any other day, at least I cannot recall anything or event that was particularly different. There was, of course, Christmas music over the PA system, religious services in the Howard Hall Theater and the regulation turkey dinner. But otherwise, everything was routine. Some may have saved a package from home to open on this day as a kind of celebration. Christmas in 1945 was in some ways even bleaker than it was on the previous two years. But at the same time, there was a pervasive enthusiasm because we were at last on our way home.

It was about the twentieth of December when the brief Paris sojourn of our band of repatriates, including my platoon mates, Foiles and Szucs, came to an end. We were departing from the AAF/ET Reinforcement Depot, Paris station in the *Bois de Bologne* forever. We were now casual soldiers and were no longer a part of any "real" Army Air Force unit. Our A and B barracks bags were packed. The bags were loaded in the back of a deuce and a half U.S. Army truck. When the bags were all loaded, we scrambled up into the back of a second truck and parked our butts on benches along the sides of the bed. Then we were off on what was to be the major portion of the French segment of our long journey home. And as usual, we did not have a notion of where the trucks were headed this morning. After a few hours journey, the truck stopped at the next way station and we disembarked from the rear. Barracks bags were recovered from the heap into which they had been tossed from the back of the other truck. We were now at Camp Lucky Strike, just one of several embarkation "stageing" camps that were named for the one of then popular brands of cigarettes. All of these cigarette camps were situated around and near the port of Le Harve. At this point,

I was about one hundred miles from Paris and still had about six thousand miles to go before I was back home again.

CAMP LUCKY STRIKE

In this modest pavilion we (Szucs, Foiles and Cook) observed our Christmas in 1945.

Still together, my old platoon mates, Foiles, Szucs and I retrieved our barracks bags from the pile and lugged them to our assigned tent. Here we were back in the U.S. Army pyramidal tents for the first time since leaving Camp Ripley, Minnesota over two years before. There were no stoves in these tents, but this was only a minor problem as the temperature was in the mid-sixties during the day, and not more than twenty degrees colder at night. But in any case, we would be gone from this camp in only a few days. The sky was a uniform gray, and sun was never seen during our short stay at Camp Lucky Strike. It was our good fortune that it did not rain during our wait at the camp. All of the concrete floored tents were connected to each other, to the down the slope latrine and up the slope mess hall, with walkways of the same material. Both the latrine and mess hall were in larger Army tents, with all of the required equipment set up on concrete floors. The PX, some offices and the dispensary were located in a few semi-cylindrical corrugated metal roofed Quonset Huts.

The repatriating troops were being staged at this and the other cigarette camps until a full shipload was amassed, processed and ready to embark on the voyage back across the Atlantic to the United States. Currency had to be exchanged; we turned in our remaining French francs and received in exchange crisp, new U.S. bank notes with a gold seal. A gold seal, the like of which was on the old gold certificates, paper currency that had been redeemable in gold and circulated before the U.S. when off of the gold standard in April of 1933. A customs declaration was required to return even though the departure from the United States had been somewhat less than voluntary and not as a tourist in any case. With all of the paper work completed there was not much else to do but wait. But by this time, we had lots and lots of experience with waiting, and we were very, very good at it. The Paris edition of *The Stars and Stripes* was available every day, and any Armed Services edition books that we might still have with us were read. Standing in the chow line and then eating our meals also expended just a little more of our anxious stay. At this camp, standing in the chow line listening to the Armed Forces radio broadcast on the PA system helped pass the time. The programs kept us somewhat amused and maybe just a little bit informed. While standing in the chow line on the twenty-first of December, we heard of the death of General Patton, who was killed in a car accident on the autobahn at Heidelberg.

With all of the processing completed, we still had a weekend and Monday to kill before Christmas on Tuesday. It may be that loading and sailing of the troop ship was delayed until after Christmas. A seasick Christmas would not be a happy holiday. Services were held in the mess tent on Christmas morning. Otherwise, it was more or less like any other day in the camp. But, of course, we were served the regulation holiday turkey dinner with all of the trimmings in our mess kits for the very last time. This was our very best Christmas in the Army, sparse as it was, because the very next day, we would be on the way again headed home.

CHAPTER 74

Going Home-Tempest Tossed

Breakfast on Wednesday, the twenty-sixth would be the very last meal that I would ever eat from my mess kit. Returning to our tent, the mess kits were stowed in one barracks bag, and then, with both bags, we loaded into the rear of an Army truck. A short time later, we were off on a quick trip to the port of Le Harve. This port had been a many times target during the war and was now more than a shambles. Separating our baggage from the heap, we fell in behind our leader and headed down the slope towards the water. Walking up a short ramp, we were on the sloping surface of the beached, first section of a floating dock. The slant of this first segment would probably be slight when the high tide came in. The long pier was made up of a number of large, gray-painted, square steel pontoons. These square tanks had been hinged side to side to make a long floating dock with several feet of freeboard. These pontoons were wide and buoyant enough that the pier could have accommodated a loaded two and a half ton Army truck, but we had to walk all the way.

Walking out on the dock, there was a bounce to our step due to the slight up, down and pitching motion of each pontoon. The large vessel that we were about to board was moored to one side of the floating pier. It was a troop ship, the gray painted USS Mount Vernon that we were subsequently informed had been the liner SS George Washington before she went off to war.

Lugging our bags, we went up the sloping gangway and into the ship through what had been the passenger-boarding doorway. Still following a leader, we walked along passageways and up stairways to our assigned compartment. Each GI picked one of the chain supported, pipe frame with laced canvas bunks and spread out his bedding. This compartment would be our billet for the next week as the ship ploughed its way across the Atlantic. Besides the bunks, this compartment was provided with something that foreshadowed times to come—a shiny, bright fifty-gallon GI Can. This can was restrained with several lengths of line to prevent its tipping over or slipping around on the deck. And all too soon more than one seasick soldier at a time put this GI Can to use as a large barf bucket.

The day dragged on as we watched more and more GIs arrive from the various staging camps and board the ship. The next morning, the twenty-seventh, chow included the traditional naval breakfast fare of baked navy beans, probably served as much for the shock value as for any other reason. A short time after breakfast, the mooring lines were cast off and hauled in. The ship got under way, backing out and around and picking up speed as she headed out of the harbor and into the open, rolling sea. It would have been advisable for most of the GIs to have—passed up this morning's chow, because most did not retain it for long. For as soon as she was clear of the harbor, the USS Mount Vernon increased speed, turned into the windy, turbulent, wave-roiled sea and commenced to pitch and roll. So, very soon, many of those who had come up on deck to watch as we started our voyage home were soon manning the rail.

A NICE DAY ON THE NORTH ALANTIC

THAT'S THE WAY THE BALL BOUNCES

The North Atlantic in mid-winter is usually not the best time of the year for a pleasure cruise such as we were on now. But here we were and so had to endure as this was the only way to get back to the U.S.A. There is an old naval tradition that on New Year's Eve, the midnight entry into the ship's log be written doggerel, such as:

> Midnight on the tempest tossed North Atlantic,
> The wretched GI passengers were nearly frantic,
> Pitching and plunging it bow into the sea.
> On board this ship was not the best place to be,
> Because as she cleave the waves, the Mount Vernon rolls and rocks,
> Bouncing the soldiers to and from on their cots,
> As the year 1945 comes to its end,
> With any luck, 1946 will see them all home once again.

And then on the seventh day of our journey, the third day of the New Year 1946, land was at last sighted. Unlike our long past departure when the troop ship stealthily sailed away from New York in the dead of night, the return was in the "bright" light of morning. The main deck was crowded shoulder to shoulder with soldiers watching as the ship slowly made its way into the harbor. Rising from the shore at a number of places facing the sea, large signs that expressed praise, gratitude and welcome to all of the returning victorious servicemen had been erected. These billboards had been placed on the shore by various so noted organizations, and if the messages did not apply, they should not be read. With our lagging return back from Europe, these tidings were somewhat obsolete. The main deck was almost entirely filled with the victorious "enlisted men" seeking a vantage point where they might catch a glimpse of New York. But at the very same time, the next deck above the main deck with the better vantage point was almost empty. That deck had been the location of the first class cabins during the ship's heyday as a luxury liner. Now, as a troop ship, that deck was officer country, and as such, was off limits to the rank and file "enlisted men" on this United States naval vessel. The fact that in only a short few days, most of us would return to real life as civilians made not the slightest difference because for now, we were still in the Army and subject to all of its rules and regulations. There will absolutely be no breaks for the lowly GI until after very the last second that he is still in the United States Army.

It had taken the better part of a day to file all of the repatriating troops into the USS Mount Vernon. Now it would take just as long to move all of the soldiers off of the ship and transport them to Army facilities. Eventually, we were off of the ship and standing on the United States of America.

NEW YORK HARBOR

Almost Back on the USA

Chapter 75

Going Home-Turbulence Tossed

Arriving at the head of the pier we were directed to get into the first of several olive drab school type buses. As soon as a last GI climbed into the bus the door closed behind him and we were on the way to, as it turned out, Camp Kilmer in New Jersey. So here we were back in the standard floor plan two-story a military barrack for the first time since induction into the Army in early March of '43. It was indeed somewhat strange that the near universal experience of living in a regulation military barrack had been eluded except for the very limits of service in the Army. But true to form it was still in the old Army game with nothing else to do but wait: wait for processing; wait for transportation; wait to be discharged; wait to go home at last; just keep on waiting.

The next day was Friday and it was entirely devoted to standing by for the call to start the processing of our dis-joinment from the United States Army Air Force. Finally on Monday we were called and it was now time for the government to take the back all of the stuff that had been provided to us on loan. The Army reclaimed all of its field equipment, bedding and most of our uniform clothing. We were allowed to keep our overcoats as well as the dress uniform that we were wearing. One barracks bag was allowed in which all of our personal things including towels, underwear and socks could be carried home. I did keep my mess kit for historical purposes as I considered it a personal item having used it to eat so many meals; and any way I knew that in due course it would end up in the scrap pile some place. For the remainder of our service in the Army we would be using bedding from the barrack supply.

CAMP KILMER

Back in authentic regulation, but war weary Army barracks.

Now divested of all the stuff which the Army wanted to keep, there was nothing else to do what the cool our heels until transportation to California was scheduled. This was the point where I parted company with Foiles and Szucs as we each would be going home a different way. Finally on Monday the seventh having been back in the United States for a complete four days, a group of GIs who are all going to the Los Angeles area were loaded onto a bus for a short trip to the airport. Where with our new reduced belongings in hand climbed up the sparse boarding stairway into an Air Transport Command C-54 Skymaster. This particular aircraft was typically configured for the primary function of hauling cargo and had the usual minimum accommodations for passengers. Each soldier after he came through the door picked one of the "bucket" seats in the aluminum benches hinged to the fuselage frames, and sat down. After buckling up one of the standard Air Force seatbelts. Our conversation was about when we would be taking off and why it was taking so long. The engines started one by one, but still we waited and waited until at last the C-54 started moving and eventually took off and headed west.

Our trip was just one more flight in the Air Transport Command **TRANSCON** project, which was devoted to ferrying returning GIs across the country from each coast to the other.

BACK IN CALIFORNIA

> Waiting around in Long Beach for our transportation, almost but not quite home.

About the only enjoyable thing about this flight was the fact that we were headed for California. The fuselage was not pressurized and so the Skymaster was flying a very low altitude. To view the ground below the airplane I had to standup and look down through one of the small oval windows. Watching the earth below as the plane flew over it I saw the source of our rough bumpy ride. I could see that the aircraft went up when it crossed over on to a plowed field and then would drop back down as it crossed onto a field covered with green vegetation. Up and down, up and down the airplane bounded along hour after hour while all of the transiting GIs were just on the verge of airsickness. After many hours of being turbulence tossed the C-54 finally let down for a landing on Love Field at Dallas, Texas. The wait was interminable, maybe a half to three quarters of an hour, until we climbed back on to our aircraft and took off on the way again in California.

The afternoon was half gone when the C-54 touched down on the runway of the airport in Long Beach, California. Interestingly this airport was also the location of the Douglas Aircraft Co. factory where in the course of events I would report to work on my very first job as a brand new mechanical engineer just short of 10 years later. Once out of the airplane and on the concrete we milled around for a while waiting for transportation to our next destination but so really glad we back in Southern California. A bus arrived and we got into it for a short ride to the separation center at the Fort Mac Arthur lower reservation in San Pedro. Apparently, with so many GIs arriving at the Fort there was a backlog in processing their discharges from the Army of the United States. Because, shortly after being checked in each of us was issued a Class A pass valid to six thirty Friday morning; but which was also restricted to a Los Angeles and its vicinity. Now, this was the one and the only overnight pass, and for that mater the only Class A pass, that I was ever issued while serving in the continental United States. Most of our group were from Los Angeles or as close suburbs and thus were able to make a short visit to their homes just prior to discharge from the Army. But, for a few members of our group it was much too far, it would cost too much and there was not enough time for the round-trip to our homes; excluding the effect that such a trip which stretch the meaning of "vicinity". So those of us who are unable to visit our homes spent three days cooling our heels in San Pedro and sleeping over at the Fort.

Friday was mostly a day of waiting around, nothing new here, with only a short period devoted to completion of the separation paperwork. We were being processed as a group and we would be discharged as a group. I was handed my decorations, a European African Middle Eastern Campaign Metal, a Army Good Conduct Medal and a World War II Victory Medal; each in the small manila envelope. The next morning, the twelfth of January 1946, after partaking of Army chow for the very last time we were paid off and handed our discharge from the Army of the United States. I received the first $100 of the three hundred dollars mustering out pay, with the remainder to be sent into monthly installments. All GIs receive the same amount to ease the now ex soldiers back into civilian life. My payoff included $1.75 travel allowance, which was probably what the fare was to return to where I was living when drafted into the Army, and then there was my accrued Army pay, up to this last day amounting to $74.18 for a grand total of $175.93. Plus I still have a few bucks in my pocket from the last payday on the first of December while still in Germany. Not a bad cash position to start a brand new life after almost three years in the U.S. Army.

THREE-DAY PASS

At last an overnight pass in the USA, just a little bit late in the game.

After the last man on our group was paid off and received his discharge the US Army provided transportation to downtown Los Angeles for us brand new civilians. While we were still in uniform how could any one ever tell that were no longer of the Army. There was a choice of depots at the end of the ride to LA, and that I opted for the Greyhound bus station as the bus would be faster and cheaper than the railroad train. A few hours later I was watching familiar but somewhat change scenery through the bus window. After making several time-consuming stops the bus departed from San Bernardino and started the last but longer leg of the trip and was soon headed along Route 66 up through the Cajon Pass. I was surprised at the major improvements to a highway that had been made during the time that I had been away. Once over the summit of the pass the Mojave Desert seemed to be unchanged. Climbing down from the bus in Victorville I had two choices for transportation for this very last but shortest part of my long journey home; hitchhiking or taxi. I was somewhat eager to get home just as fast as I was able, so hang the expense I hired a taxi and off we went.

When last I had traveled the way from Adelanto to Victorville the trip was about eight miles long. But, unknown to me, while I was gone, the distance between the two places had become shorter. A new section of road had been cut across the desert from Route 66 to the George Army Air Force Base. This new section of highway bypass the section of the old road which dropped down into the Mojave River Valley, wandered around a little and then climbed on a long straight grade to exit the valley and head for Adelanto. This all-new section of roadway with its cuts and fills headed up a slight grade directly towards the airbase. This was a good road and I was pleased to see that the Army pay that I had invested in war bonds had been put to good use.

When we hit Highway 395 the taxi turn right and I said to the driver, "stay on Adelanto Road and turn left at Lee Street. I'll tell you where to stop." While the streets that had some homes or other developments on them had long since been paved with asphalt, there were still no house number street addresses in Adelanto. It had always been somewhat of a social event—to go downtown (such as it was) to visit the post office in the late afternoon and pull the mail from your combination lock PO Box. And of course it was a real social event when I said to the driver "this is my home, turn into the next place on your right." He turned the taxi in the sandy driveway and stopped. No one including myself ever knew just when I would finally return from the war but this was a grand homecoming as my mother Ruth rushed out of our home and greeted me with a long tight hug.

PART
VI
BACK HOME AGAIN

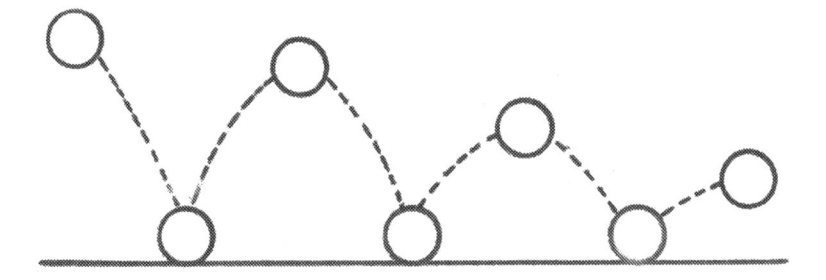

Chapter 76

Lost in the Desert

The end of the war with Japan was almost five months past [75] and all of the home front trappings of the war had long since vanished from view when at last I had returned home. Missing from the windows of so many homes were the small rectangular red-bordered banners blazoned with blue stars for those serving who were expected to come home sometime and gold stars for those who would never ever return. Never once had I ever seen one of these banners displayed in any window as all of my off duty time in the United States had been spent in either downtown Miami Beach or downtown Little Falls. Probably relegated to the trash were all of the many maps of the world on which, such as my sister later told me, the progress of the war had been was plotted and the known locations of friends and family were posted. At the time none of this was of great interest to me as was the importance of just finally being home.

I was at a second starting point much like being fresh out of high school and just figuring out what to do with my life. This new world was quite different and somewhat stranger than the prewar one and decidedly different than the last three years of life in the Army. Except for the new roads the desert appeared to not have changed, but many of the people who I knew were changed. The increase in the height of the town's children, most of all my little brother and sister,

---

[75] This was starting to cut it close because when drafted the maximum length of service was to be for the Duration of the conflict, plus six months. As it turned out this was somewhat fuzzy because the "duration" was not demarked by the surrender of Japan, but instead by the President's proclamation that the war had ended.

who were so much smaller when I went away, was astonishing. I was really not a part of this new present and somehow needed to figure out how to fit into this new way of life. This strange new existence was so different the last three years of my life that it would take some effort and some time to adjust. The world had changed as I had also changed; but the world and I had changed in dissimilar ways.

This new life was different, an extreme switch from that to which I just bid farewell. I still remembered the 1257th MP Company; what a good and grand group we had been, true comrades of boredom, misery and mirth. We had performed a first rate and maybe even important job that the fate of war had assigned to us. But now the 1257th was over and gone and never would be again. I was so very glad that it was all over with and yet at the same time I missed what we were and what we had been. It had not been too bad of an experience; an episode to always remember, but not one to ever want to relive again.

Now twenty-one years old, which in California meant that I was now "legally" mature enough to buy and to consume all types of alcoholic beverages and to vote in all elections. Still only twenty-one-years old I was technically still subject to being drafted into the armed forces by the selective service. So by law I was required to register for the draft and to carry a draft card on my person all times. The draft classification this time around was 4A-Service Completed. If my service was completed then why was I required to resign up for the draft instead of being presented with a card that stated that I had served long enough. Having to carry a draft card at all times rankled me then and, the idea, still rankles me to this day.

On the morning of my first full day home it was time to go and see what if any thing at the center of Adelanto had "changed" during the time that I had been gone. Also, and just as important I needed to find a ride to Victorville. I wanted to go to Victorville for two very good reasons. First, I had to make the trip into the town to buy some new civilian clothing because all I had that to wear was one army uniform. And second, I wanted to find out if my buddy Vernon and for that matter if any other of my friends were in the town or somewhere else to in the area. Setting out from home, I crossed the street and walked through the desert on the sandy shortcut to the center of town. This was the same path that I had walked with my father as he led the way for my first day of school more than sixteen years before. It was good to be taking this path again now that I was back home. During my time away, so far as I could remember, the center of the town such as it was had changed little. The grocery store had a new proprietor. And of course the adjoining combined automobile repair garage and service station was no longer the business of my Aunt Eleanor and Uncle Elias. He was now employed at the George Air Force Base water system pumping station on the Mojave River. They had moved from the living quarters attached to the back wall of the garage and into Government furnished housing. Brand new homes had been constructed along both sides of the street running from what was then Air Base Road to the main gate of the air base. I would soon visit my Aunt Eleanor and Cousin Carol Jean in one of these houses but on this day I had to go to Victorville.

My first stop after arriving in the town was a clothing store to acquire a minimum new civilian wardrobe. A major article of apparel purchased was a powder blue sport coat (probably because it was on sale); what did I know or care about current fashions. A sport coat, the wearing of which, in time would precluded the admission in to the Hollywood Palladium. This was a venue where, in those halcyon days, proper dress and decorum required that a suit coat (no exceptions) be worn to see and hear to a big swing band. Hanging on to the bag of brand new attire I set off up to grade walking on the sidewalks where they were, but for the most part just crunching along on the sand and gravel. The town as far as I could see, was not much changed from the last time that I had walked this way. Victorville was still more or less confined to the steeply sloping area between the slightly sloping brush covered sandy desert and the Cottonwood tree bordered shallow Mojave River. I was on my way to Vernon's home to find out if he was in town, visit with his mother for a while and to maybe get a ride back to Adelanto. As it turned out he was in town and so we had a long catch up conversation. Late in the afternoon he said that he would give me a ride home on his way to work. So off we went in his rattletrap car passing by the Air Force Base, to which he would shortly return for his job as a security guard.

There is always some trauma resulting from any change and getting out of the Army it was a very preeminent change to my life. So, having been immersed in military culture for almost three years, an eighth in my life at this point, getting used to this new situation free of structure and direction was going to take some time. Enjoying the new freedom of being able to go wherever I wanted and do just what I pleased I was in no hurry to find employment. I thought that I deserved a vacation, this was really the first free time that I had since my induction into the Army. The weeks furlough in England had not really been much of a vacation. And soon my mother was on my case about getting a job while all I really wanted to do was enjoy my new freedom for a little while longer. And besides I had an income of one hundred dollars for each of the next two months.

Also, if I were to find a job the first thing needed was some form of transportation to take me to wherever I would be employed. Small and somewhat isolated Adelanto did not have any real public transportation. So what was needed was an automobile that I could drive to work. There was someone in the town who that a very use small 1942 Ford V-8 for sale cheap. With only a small down payment, the rest due later, after I found a job. I now had "wheels" but as yet did not know how to drive a car. So, when the car was delivered to our home I set out to teach myself how to drive. Practicing on the fortuitously usually empty streets of Adelanto I had in a short time mastered the fundamentals of driving and was able to do a fair job of getting to where I wanted to go. After parking the car in our driveway my brother pointed out that I had been driving on the wrong side of the road when I came home. Of course driving on the left side of the road did seem perfectly normal to me having just spent over two years in England. But in any case I had gained sufficient skill to drive around locally and in all probability would at some time obtained a driver license.[76]

For all of the unemployed unskilled labor—that is those servicemen who taken up arms more or less fresh out of high school—the quiescent George Air Force Base was the only venue that offered reasonably paid work, which was as a civil service security guard. This Army Air Force Base was no longer operational and now was being used for storage of B-29 Bombers back from the war with Japan and two engine training aircraft which more than likely had been used at this very Air Station. The military presence was minimal, but there

---

[76] I finely obtained a driver license by writing to the California Department of Motor Vehicles for one while serving in the U. S. Navy Atlantic Fleet. So, I never did take the driver test to obtain my first driver license.

was a significant civilian workforce employed in the maintenance of the parked aircraft. The security function consisted of two different tasks. One duty was the around-the-clock control of access through the base main gate. The other responsibility, the roving patrols which were limited to the hours when the maintenance workers had all gone home. On the two night shifts the different the assignments were the as follows: security at the main gate; solo patrol in the weapons carrier along the unpaved roads which meandered through the desert past all of the parked aircraft; or solo walking patrol through and around the various aircraft hangars and maintenance shops while carrying a night watchman's clock.

The security force was made up of fresh and young veterans who had served in all of the branches of the military service and in all of the theaters of operation. Probably the most interesting member of the crew was a former Marine who had been that part of that band known as Carson's Raiders (a special forces unit); who booked a certain amount of stature with the rest of the crew based on this fact. A crazy kind of a guy who road a motorcycle and who was frequently being forced off of the road by cars. Probably he was trying to go places where there was more than likely just not enough space. Who knows how long he survived in civilian life with this problem of judgment.

When finally I made my mother happy and hired on as a security patrolman there were already two other members of the class of '42 working at the airbase. There was my pal Vernon, who had told me about this job, and Chuck Britten. In three of us had also been some time members of the Mighty Eighth Air Force in England. Britain had been a pilot and Vernon was a waist gunner on B-24 Bombers. Now here we all were on the same job in this "hometown" Army Air Force Base security unit. Of us three my particular military experience had been the most apropos for our present job, such as it was. In any case working as a security guard, a function with which I was already somewhat familiar, unequivocally held no allure as life's occupation.

Perhaps because at the time it was a government job somewhat directed to providing work for unemployed ex-servicemen every man on the security force was issued a Colt .45 automatic, holster and web belt. This Side Arm was the very same kind of weapon which I had worn strap around my waist during and for many months after the Second World War. We had all been trained in the use of firearms to shoot people as part of our recent line of work, but even so one day our assignment was to report to the firing range for target practice. This training session was somewhat ironic because I shot off more .45-caliber rounds on that day in than I had officially fired while the war was in progress. This training session did reinforce the skill with and the weapon that I had acquired during the surreptitious target practice in Germany after the war was well over.

On one warm and moonless night he was my assignment to drive patrol duty out among all of the scattered conserved aircraft. Solo in a weapons carrier with its top down, possibly to provide a clear field of view, I followed the sandy road as it wound through the desert vegetation and passed all of the preserved airplanes. The Police style visor hat lifted and started to blow off of my head so reflexively I reached up with one hand to hold it on. This was a mistake because in the process of grabbing my hat the one corner of the vehicle's windshield cut a gash in the underside of the low wing on one of the two engine aircraft. When I when off duty at the end my shift I reported the accident, drove home and went to bed. Early the next morning the chief of the airbase security and one of his lieutenants arrived at our home. He said that I had been driving too fast (and maybe I had) when I hit the aircraft, and so he discharged me from the security force. He retrieved the Colt .45 automatic and told me that I was entitled to a hearing if I wanted to contest being fired. But considering the fact that I did not as yet have a drivers' license, (of which they were not aware) it was probable that a hearing would not result in a finding in my favor so I decline.

So there I was out of work; this job such as it had been was now ended. This employment had been the source of income but otherwise it was not very interesting or for that matter not that much different from being in the Army. But still needing some income I found some odd jobs and then somewhat steady work with the chicken and egg farmer; about the same kind of work I was doing after hours while I attended high school. At this point I could have applied for the 52-20 unemployment benefits, but I would have to go down to San Bernardino to sign up. The U.S. Congress had some how realized that there might be a few ex servicemen who temporarily might not find work and so had institute a system of unemployment benefits. This system of compensation would pay the unemployed recent members of the military $20 per week for one year (52 weeks). There were some few veterans who took advantage of the system and remained happily unemployed for a whole year. And then there were the recipients of 52-20 who had been quickly ejected from military service when the war was over and who not so much as not being able to get a job had some difficulty adjusting too the regular employment of civilian life. A way of life to which they had become accustomed ended very abruptly when they were dumped and they did not have an established prewar life that they could return to and continue.

One night when I drove to Victorville to shoot a little snooker with Vernon, he told me about a program where if we took the Captain Eddie [77] Test and received a satisfactory score we would be excepted to enter directly into the U.S. Navy Electronics Technician Service School. This naval service school was accredited for college credit and so this would be a start on a real technical education. It sounded good to me so I said to him that we should go take this test because at present we were going nowhere career wise. A few days later we drove down to San Bernardino and took the Captain Eddie Test. A week or so later we each received a notice that we had passed the test and would now have to enlist in the United States Navy to attend the school. After deciding to sign up we went down to the recruiting office in San Bernardino to enlist. After passing a physical examination and signing the enlistment papers we were given a few days to take care of our affairs and then would be on the way he to the Great Lakes Naval Training Center near Chicago.

On our last day in town we were walking up the east side of Victorville's main drag on sort of a farewell final visit. Heading up the grade, nearing the high school, we met Chuck Britten who was coming towards us and so we stopped to talk. We told him that we both had enlisted in the Navy and be and would be on the way to Chicago the next day to attend the electronics technician school. Chuck said that he could relate it to what we were doing as he was

---

[77] Captain Eddie had been responsible for developing a test of general math and science ability to pick personal for admission to the U.S. Naval Electronic Technician Training School.

in the process of returning to the Army Air Force. He would make the Air Force a career and retire from it as a full colonel.

So there were at least three of from the class of '42 who having some problems with adjusting to civilian life decided to return to the military life. This deed was duplicated many times in my particular age group, those who were fresh out of high school when they went into military service. Over the years I have been met many and heard of many more from this age bracket who had also returned to the military, if only for one more time. Curiously very many of those who went back into the service, chose a different branch of the military than that in which they had served in the first time around.

PART
VII
THE DENOUEMENT

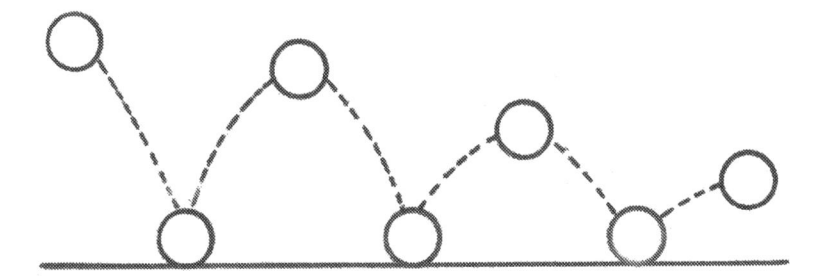

CHAPTER 77

The American Soldier in World War II

The official Department of Defense U.S. Army World War II personnel statistics are enumerated in *The World Almanac*. The numbers listed are for the five plus years between December 7, 1941 and December 31, 1946. While the conflict in the PTO had ended on August 14, 1945 (VJ Day) the war was not officially completed until President Truman had on the final day of 1946 issued a proclamation on the formal cessation of World War II. So there is some fuzziness as to how many men and women actually served in the US Army during the period in which combat was underway. It may well be impossible to determine the actual numbers however the official ones are as follows:

CATEGORY	COUNT
Number who served	11,260,000
Battle deaths	234,874
Other deaths	83,400
Total deaths	318,274
Wounds not mortal	565,861

The number of those who served has plainly been rounded off. Probably this number was rounded up, but in any case is more than likely within a half of a percent of the correct value. The duration of individual service is another matter entirely. Some who served were in the Army when the United States suddenly joined the war and were still in it when the war was officially finished. While at the other extreme some served only after all the fighting was finished and some few serve for only a day for various reasons. But the peak number during the actual conflict was probably close to the total number who served during the recognized limits of the war. Overall about eleven percent of the average wartime population of the United States served at some time in one branch of military service. Almost everybody had a relative or at the very least knew someone who was serving during the war.

The tally of battle deaths is a precise value for another fuzzy summation. Battle deaths are multi-form; while the most were the direct result of enemy weapons and munitions many were due to other causes. But not even all of the battle deaths, which resulted from enemy weapons or munitions, occurred in close combat before the enemy. Anywhere in a war zone there is always the chance of becoming the random recipient of a bomb, an artillery projectile or a missile. There were deaths due to training events such as the debacles of Slapton Sands where two LSTs (Landing Ship Tank) were sunk by German torpedo boats and then when some troops were landed ahead of schedule may have been hit with friendly fire. There is always a chance of receiving friendly fire or having an accident in the confusion of battle. Accidents due to failure of judgment by weary and distracted soldiers and accidents due to mechanical failure of equipment and weapons did result in deaths. Other fatalities were the result of murder, suicide, medical events such as heart attack and stroke—senior commission and noncommissioned officers for the most part—, freezing, flame or poisoning—bad food, bad hooch-. Some would succumb to wounds of war long after the battle had ended. Distractions can sometimes lead to errors in judgment with deadly results. One such almost event related by Alvin "Bud" Anderson concerned the attack on a German railroad train. Piloting his P-61 Black Widow through the dark of night hunting for targets of opportunity he spotted a blacked out railroad train highballing along the rails shining brightly in the engine's beaming headlight. Flying and firing head-on as the airplane closed in on the oncoming engine he fixated on the shaft of light boring through the pitch black night, the deer in headlights syndrome. He said that he would have flown right into the engine had not his radar man excitedly yelled at him "pull up, pull up!" Flying right into a railroad engines bright headlight, Bud said, was probably what had happened to several P-61 night fighters that were mysteriously lost by his squadron.

Excluding the deaths due to enemy equipment and munitions just about every other cause discussed above is likely included in the precise count of "other deaths." Additional causes such as training accidents, including friendly fire, time acting maladies and some times even old age are more than likely included in the accounting of other deaths

The array of causes of the "wounds not mortal" was much the same as the source of battlefield deaths. That is the majority of the wounds, including in conspicuous psychological problems, were due to enemy equipment and munitions. For very many veterans the lingering disabilities, afflictions and pains, including uncountable instances of mental distress, would last for a lifetime. Yet some wounds were only minor with little lasting effects or problems, and maybe some wounds could even be classified as trivial. There was a story, probably apocryphal, of an infantryman who when on the line discovered that he had the Big Big-G. So he was sent back to the aid station and was listed along with the real casualties and in due course was awarded a Purple Heart.

What these official Department of Defense statistics fail to enumerate is a number of soldiers who were missing in action and otherwise[78], or the number who serve some part of the conflict as prisoners of war. Probably no one cares to know how many GIs were suicides or given dishonorable and bad conduct discharges, or for that matter how many soldiers served hard time in the Army penal system or who were in fact executed for crimes both military and civilian

Nor is there any breakdown given, or for that matter any way to know just how many soldiers in the grand total who served were actually involved in direct combat with the enemy at one time or the other. There is a somewhat common idea that all of those who served in the Army were it some time or the other involved in direct combat. Freeman in *The Mighty Eighth War Manual,* states that eight soldiers were necessary to back up each airmen who was flying. That may have been true for the highly technical bombing operations in England. However, the general understanding is that on the average two GIs were required to back up each one who was directly engaged

---

[78] Some of those who were part of "the missing" were only over the hill. It has been estimated that there were probably at least ten thousand deserters in Paris alone, missing but not out of action.

with the enemy. So there you have it, there were twice as many noncombatants as there were fighters in the Army notwithstanding that all were ready to engage the enemy if called upon.

In so far as engaging the enemy and there are always those who delight in war, the hazard, the violence and the release from civil constraints. That there are members of the human race who exalt in the role of warrior is not surprising. It is part of mankind's nature to envision the infliction of misery, suffering and even death on some fellow members of society. While the majority may decry a war as a matter of social principle, deep down in their heart of hearts they sometimes welcome war. Many find an aspect of excitement in war, especially if it is only experienced vicariously; *"our boys are now in action"*. And perhaps those who gather to demonstrate and denounce war may protest too much.

JOHNSON   REARDON   VERNON

Had not Al Reardon enlisted in the U.S. Army during December of our senior year there would have been twenty-one boys, seventeen and eighteen years old, in the graduating class of '42. So far as it is known all twenty of the remaining male members of the class spent some time in one of the branches of the armed forces and survived the war. It was matter-of-fact our unquestioned duty to serve our country, no doubt as it was also for our counterparts of the "hated" enemy to serve their countries.

In our high school class there were four of us, Merl Johnson, Al Reardon, Bob Vernon and myself who became best friends because of our shared interest in science. In this day we would probably be called nerds, but this was in the time before nerds and besides our high school was much too small, with only two hundred students who all know each other, to meet the requirements for having nerds. Reardon was stationed for some time on Hawaii and then on Oahu. The troopship, that was transporting his unit, grounded on a reef as it sailed from the Marshall Islands on the way to the invasion of Saipan. Arriving late to Saipan he was involved in the defense directed against Japanese aircraft. Vernon as you know served a tour in the Mighty Eighth Air Force as a waist gunner on a B-24 Bomber. Johnson, whose ambition was to become a chemist, ended up in a four point two chemical mortar unit. This assignment may have been fore shadowed by the high school incident of the evacuation of the science classroom. His experiment with chemical warfare was implemented by dropping a capsule filled with solid tear gas on to a lit Bunsen burner. This, tear gas filled, capsule had been provided to him by a classmate, who's father just happened to be the Commanding Officer of the George Air Force Base. He was wounded during the Battle of the Bulge in the Ardennes's forest of Belgium. As a consequence of his injury Johnson was awarded compensation for one hundred percent service-connected disability. The Army's plastic surgeons did a fine job of rebuilding his face, but his glass eye always had a kind of a wild-eyed look.

The circumstance that only about one third of those who served in the Armed Forces were directly involved in combat may be why, even among ourselves, we were disinclined to talk about our war experiences. It may be that no one talked about their role in the war less anyone else think that they thought that their service was more or less important than anyone else's service. Occasionally some apropos, comment might be made during conversation such as the time Vernon and I had seen the motion picture *Twelve O'clock High*. While I was somewhat offended by the atypical depiction of the

MPs in the general court-martial segment, Vernon noted that the scene where a B-17 Bomber returning from a mission crash-lands, explodes and burns was pure Hollywood hyperbole. He observed that sometimes bombers had very well crash-landed when they returned from a mission; but they never exploded and burned. There might have been as much as two hundred gallons total remaining in all of the fuel tanks, Vernon said, which was not enough for any kind of an explosion or a fire. The two of us had roomed together while we attended college on the GI Bill, and during this time almost never said anything about anything that had occurred during our service in the Army. On one occasion he did tell me a somewhat funny event and thus an admissible story. There was an exhaust driven supercharger at the top of each of the B-24 Bombers four aircraft engines. Sometimes gasoline would leak into the engine exhaust and catch fire in the turbine blades, in which case it was known as a "torching turbo". On one flight the navigator informed the pilot that there was a torching turbo on one of the engines. To which the pilot replied, "what you want me to do, piss on it and put it out?" Such were the things that we could tell each other.

So, in that time we did not know or really care, as it was not important what each of us had done in this latest Great War. We all had been on the same team and had performed as best as we could in the position to which we had been assigned, and to us in the total scheme of things no one was more important than anyone else when it came to attending the war. But it was, as we all knew, that some positions were more dangerous than others were. Certainly we all had been tempered by our shared adversity of the Great Depression. It had been a time when everyone, to at least some extent, had been afflicted by our common troubles. Also, significantly the particular historical circumstances of the foundation and the evolvement of the United States influenced our attitude, a new and different civil attitude that binds us together. A civil attitude ensuing from the fact that America is a somewhat new and unique nation-state in the entire history of the world. Therefore, our outlook towards war, based on military traditions arising from past wars and in particular the Civil War, is also different from that of other nations. Thus the aerial assault upon Germany was a task that was to be preformed with an assured élan, as exemplified in the conclusions of (Freeman) *The Mighty Eighth*:

> Untempered by ideas of caution, the men of the Eighth Air Force pursued the concept of strategic bombardment with dogmatic faith. It was this fervor to get things done, to surmount all technical and operational obstacles that took the Eighth further along the road it chose than ever the British or Germans would have deemed possible. Procrastination and the negative were scorned, and even bloody experiences did not deter the over whelming intention to succeed.

And through it all at heart, in the back of most of our minds, we were first and always free citizens of the United States of America and so were not totally committed to the Army way of life. But still being team players we did our duty in our own attentive way; instead of the harsh blind duty of those who served for the totalitarian states.

Also, there was the example of our forebears, the veterans in the first phase of the world conflict, the war to end all wars. The whole country was always somewhat amused by the antics of those veterans in the parades that came with the American Legion national conventions up to the United States involvement in the Second World War. There was little doubt that World War I had been the seminal event in the life of these veterans. As a boy I knew this when I saw the happy camaraderie of my father and his fellow veterans. Take nothing away from what these Veterans of World War I had accomplished; but the fact is, the United States had joined that war for only its very last fourteen months. In contrast fourteen months was only getting started in World War II. A case in point is that the United States had been entrained in the Second World War for fifteen months when I was inducted into the Army. This second conflict had lasted so long and we had served so long that there was a feeling of relief when it was finally all over. We were proud of what had been accomplished but not that sure that the world had become a better place in which to live. We were more subdued and more serious group, than the self-certain veterans of World War I, and held no illusions that we were the greatest.

We, the class of '42 along with all of our close contemporaries were in the tag end of that aggregation on which had been bestowed the honorific "The Greatest Generation". And if in truth the all of us were the greatest generation it was that, annealed by the Great

Depression, we were the last truly free citizens of the Republic with all the responsibilities and risks which being that entailed. We were free citizens who performed our duty, because we needed to and because we wanted to. So, being free citizens we held a suspicion of and a well-developed disdain for **authority**. (A civil inclination stemming from the tenets of the Declaration of Independence.)— And had not General Patton himself, on the first of August 1944, requested permission to advance and then not bothering to await a reply took off with the Third Army—. This low regard for **authority** is illustrated by a story that we told ourselves in England about ourselves in England.

00000

Every day the breakers reach further and further up the sands on the near shore as the southern zone of the island slowly depresses under the burden of the men and matériel massing for the impending excursion to the far shore. The fog has congealed over the land on this pitch-black moonless night. It is a cold English fog so thick and so wet that it obscures the very existence of the land and even impedes the very act of breathing. A lone sentry is standing on duty at his post. The sound of boots crunching on gravel punches through the obscuring fog.

"Halt, who's there?"

"A Polish airman!"

"Advanced and be recognized Polish airman."

The Sentinel switches on his torch and holds the proffered papers up close to his eyes. Checking the credentials he finds them proper and in order. He returns them and says, "Pass on Polish airman."

A little while later the screeching of hobnails on paving stones pierces the sodden fog.

"Halt, who's there?"

"A British soldier!"

"Advance and be recognized British soldier."

The picket reads the soldier's papers and finding them correct says, "Pass on British soldier."

Somewhat later the slapping of boots on the paving stones penetrates the dripping fog.

"Halt, who's there?"

"A Canadian sailor!"

"Advance and be recognized Canadian sailor."

The Sentinel examines the sailor's documents and finding them in order says, "Pass on Canadian sailor."

Still a little later the grind of shoes scuffling through the gravel punctures the soaking fog.

"Halt, who's there?"

"Go f—k yourself, you stupid son of a b—h!"

"Pass on American soldier!"

—THE END—

END NOTES

1. Pup Tents
   (Bacon) *Sinews of War,* provides the information about the U.S. Army pup tents origin in the American Civil War. However, the source of this information about the pup tent is not identified.

2. White Gas
   White gas was unleaded gasoline without any additives what so ever. White gas was used extensively in the home as fuel in gasoline fired appliances (mainly Coleman) such as lanterns, stoves, ranges and space heaters. This fuel was usually purchased at a filling station. The buyers can, 1, 2 or 5 gallons, would be filled from a 50-gallon drum. And sometimes would be pumped from a larger tank, which were usually located at the side or at the back of the establishment.
   The unleaded gasoline was called white gas because it had not been dyed some color. The fuels sold for use in automobiles were all color-coded. Leaded gasoline contained tetraethyl lead, an anti-knock agent that increased the effective octane rating. Before universal electrification gasoline was pumped manually from the underground storage tanks. The gas "pump" was filled by wobbling back and forth a ball handle topped lever that operated the actual pump. Each swing of the lever along side of the tall gas station pump would provide a noticeable increase of the liquid level in the glass cylinder tank. Protected by a diamond patterned expanded metal mesh, the glass holding tank was placed high enough so that the fuel would flow by gravity into a gas tank of any car, even a Model A Ford. On the inside of the glass walled tank there was an overflow standpipe to limit the amount of gasoline to just 10 gallons. Also, on the inside wall of the tank the markings on a metal strip measured the amount of fuel dispensed in the whole and in fractions of a gallon. In the side-by-side filling station pumps the different grades of gasoline such as regular, Ethel and Hi-Test Ethel were identified by their colors. The different grades of fuel

were transparent vibrant almost incandescent shades and hues of red, green and blue depending on which company supplied the gasoline.

3. Disposition of the Repple Depple

While each airfield in East Anglia was dispersed in a number of close but separate sites for tactical reasons and to minimize disruption of the countryside, the Repple Depple was also disbursed far and wide in to many scattered sites just to utilize the existing facilities in which it could preformed its various functions. The following account, which is not complete, is based on information from my special orders, and additional information from the Repple Depple publications *The Yarnfield Yank* and the *Ardee News (*The Smithsonian National Air and Space Museum also holds these Repple Depple documents in its archives) and from (Freeman) *The Mighty Eighth War Manual.*

There were several changes in the unit designation of the command during the time that I was station at the Repple Depple. When the 1257th arrived there at the end of August 1943, the unit designation was the 12th Replacement Control Depot. Around the first of 1944 the unit designation was changed to the Eighth Air Force Replacement Depot. Later that year the unit was transferred to the USSTAF Air Service Command and re-designated the 70th Replacement Depot. Then for what I presume was reasons of political correctness (and thus I think take some of the sting from the high rate of combat casualties), the unit was re-designated the 70th Reinforcement Depot. After the war had ended and there was a change of mission so the designation of the unit was revised to AAF/ET Reinforcement Depot (PRO) (Main). Still with all of the changes of unit designations the one constant sobriquet was "Repple Depple."

There were, at the peak, seven halls located in the English Midlands with the primary function of processing the movement of Air Force personnel; air crewmen, ground units and individual "casual" soldiers to specific Air Force stations. Also those airman who had completed a tour of combat missions, were processed to provide their transportation back to the United States. After the liberation of France stations were established at Paris. Another important function of the Repple Depple was the operation of combat rest homes to provide R&R, as a respite from combat, for the aircrews. These rest homes were, with the exception of one hotel, all in fact English country estates.

Some details of the various personnel processing stations of Repple Depple command are as follows:

Sta. No.	STATION NAME	STATION LOCATION	Closed
594	Duncan Hall (1)	Stone-Yarnfield, Staffs	15Nov45
594	Howard Hall	Stone-Yarnfield, Staffs	1Nov45
594	Beatty Hall	Stone-Yarnfield, Staffs	20Oct45
594	Nelson Hall	Stone-Millmeece, Staffs	15Oct45
594	Seighford (2)	Seighford Airfield, Staffs	———
594	Stone Baggage	RR Station Stone, Staffs	———
591	Washington Hall	Chorley-Euxton, Lancs	10Oct45
591	Chorley Baggage	Euxton, Lancs	———
579	Bruche Hall	Burtonwood-Padegate, Lancs	15Oct45
569	Adams Hall	Bamber Bridge, Lancs	5Oct45
372	Orley Field (2)	Orley Field, Le Bourget, France	———
372	Chateau Rothshield	Bois De Belogne, Paris, France	(3)

Notes: 1. RD headquarters up until 15Nov45. RD headquarters reestablished on 20Nov45 Station 385, at Camp San Francisco, Chateau Thiery, France.
2. Unit located at the airfield to service the personal passing through.
3. Still in operation late December 1945 when I passed through.

The information about the combat rest homes is somewhat sketchy. Of the sixteen combat rest homes fourteen are listed in *The Mighty Eighth War Manual* as follows: Stanbridge Earls, Palace Hotel Southport, Moulsford Manor & Bucklands, Combe House, Walhampton House, Aylesfield House, Roke Manor, Pangborne House, Spechley Park, Furz Down House, Eysham Hall, Keythorpe Hall, Ebrington Manor and Knightayes Court. Photographs of two of the country estates, Stanebridge Earls and Moulsford Manor, are also shown in this history. *The Yarnfield Yank* contains photographs of Combe House, Henley-on-Thames (which is not included in the preceding list), Pangborne House Eynsham Hall and Aylesfield House. Also, a photograph of Walhampton House is shown in (Bowman) *The Wild Blue Yonder.*

MILTON COOK

The following article from the *Ardee News,* Volume 1 Number 5 provides a summary of the activities performed by the Repple Depple during World War II and some of the repatriation activities that occurred after the war had ended.

IMPRESSIVE FIGURES IN COMMAND RECORD

BY DAVE CLARK
Ardee News Staff Writer

For months and years to come we will read about records set during our world war which have been an under-cover subject while our enemies were active. The members of any supply or processing command such as ours have worked at pretty thankless tasks for many months. And, although no one would belittle the immeasurable sacrifices of the combat troops, it seems fair to claim a little credit for our work now that the surge of man and steel has reached flood peak and is flowing in reverse through our hands.

Inclusion herein of all the statistics available would fill many issues of the **ARDEE NEWS** and none would wade through them anyway. Here instead are some of the highlights of our activities.

The second half of 1942 saw a comparative trickle of 1,761 ground personnel processed through the command. We were operating a only a few UK camps in December with one Rest Home which opened in November to welcome the first 13 battle—weary veterans. In July 1943, the first combat crews arrived. The ground personnel hit their peak in December of that year when we counted a procession of 19, 121. The total number of man processed that year was 71, 167 plus 2,297 Rest Home guests. The year 1944 saw three records established and business generally humming. Combat crews hit their peak in July when we tallied 12, 957 crewmen and the year was our biggest so far with a processed total of 202, 054 plus 19, 208 guests in the Rest Homes which had increase in number to 16.

The happiest record of the year was the number of vets returning home, 10, 141 during October alone and a total of 48, 516 for the year. The first months of 1945 saw the final surge of combat crews and their disappearance from our scene in June. In March with a billeting capacity of 19, 000 we really burned up our previous records when 34, 264 new faces came and went through our five UK stations and the two newly added Continental units. So far this year we have sent 72, 063 ETO balmy souls back to the land of milk and Buicks and we are rolling up our sleeves for the months to come as predicted in Gene Emunson's story in column three. In the line of grand totals we have introduced 130, 905 combat crew members and 114, 213 ground personnel to the wonders of "Oui Oui" and "Blimey," and torn 122, 847 kicking and squalling soldiers from their homes in the ETO. All this moving, plus additional smaller categories, totals 4 20, 743, with the Rest Homes handling an additional 37, 997 since June, 1942.

At times when we stared through sleep craving eyes at heaps of forms and orders, moving tones of humanity and baggage at all hours, or answering the same questions a thousand times a day, this seemed a poor way to fight a war. But we stuck to the job. There could be no higher goal for any group of people anywhere working together then that they do their job with speed, accuracy and a sense of humor.

00000

Some history of the Repple Depple is described under the heading **Replacement Control Depots** in *The Mighty Eighth War Manual.* It is apparent from the changes in the RD unit designations listed above that Freeman's description is not complete. However there is some of the information provided in the list of **Other Stations** which does raise some questions. Aside from some questionable dates this information may be correct for the early on RD configuration. If so things had changed by the time that we had arrived at the Repple Depple. The three sites at Stone-Yarnfield are listed as follows: Beatty Hall (STA 518), Duncan Hall (STA 509) and Jefferson Hall (STA 594). When the 1257th arrived all three sites were included in Station 594. And the site named "Jefferson Hall" was in fact Howard Hall. This is a mystery, it may be that a change in this site name was used for a while and then dropped in favor of the original name Howard Hall. The two halls named for American Presidents Washington and Adams were both in Lancashire, so that it is reasonable to expect that was where "Jefferson Hall" would also have been located, if an fact it really ever existed.

There are two items in (Reynolds) *Rich Relations that* touch on the Repple Depple. First, Adams Hall is noted to be the location of the headquarters of the 1511th Quartermaster Truck Regiment (AVN) 8th AF Service Command until at least 27 August 1943. *The Mighty Eighth War*

Manual indicates that there was a detachment from Stone at Bamber Bridge in September 1942. And also that it was a QM Depot until January 1944. So it is probable that both Army Air Force units shared of the station for some time. Second, I believe that the July 1943 date, cited in *Rich Relations* as the arrival date of the WAC unit at Stone is questionable. My recollection is that there was not a WAC unit stationed in the Howard Hall when the 1257th arrived at the Repple Depple late August 1943. Also, the account of personnel processing and the build up of personnel at the Repple Depple provided in *The Mighty Eighth War Manual,* makes it doubtful that the WAC unit could have arrived in the middle of 1943. I think that the correct year was 1944 and most likely a typo occurred in the paperwork somewhere along the way.

4. Correspondence from Ellen Hurlstone of Eccleshall

    Postcards from Stafford—Ellen Hurlstone sent my mother a set of seven postcards that displayed scenes of Stafford, Staffs. Both sides of the two cards with Hurlstone's comments are shown below. There is some mystery about this set of postcards. They are good quality postcards for the use of tourists; so were most likely printed before the beginning of World War II in September of 1939. The scenes of Stafford on these cards are definitely of prewar vintage. The postcards that were generally available during the war for "tourists"—military personnel—were poor quality printing on gray recycled paper card stock. Something that is interesting about these postcards is the "sayings" printed at the top of the space for correspondence. In addition to those on the cards shown here there are some different sayings by the prime minister on other cards: *"Let us all strive without failing in faith or in duty"* and *"This is the time for everyone to stand together, and hold firm."* Now another mystery, do these sayings referred to the "Slump" or to World War II early on?

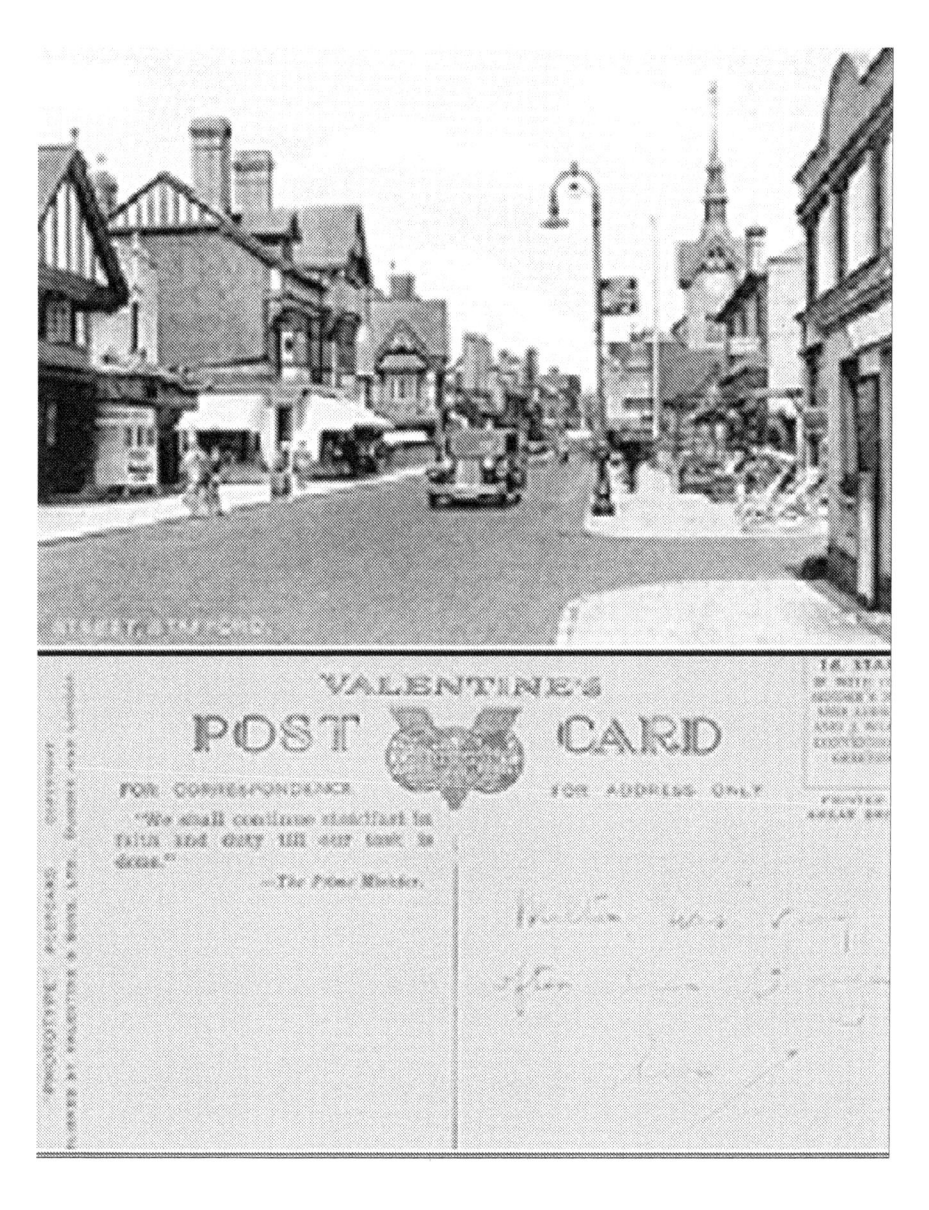

Comment—"Milton has very often been through here."

Comment—"He has walked down here many times."

POST CARDS OF STAFFORD

Ella Hurlstone sent this letter to my sister Joy in 1945. This letter provides a different prospective and window in to this different time and place. It is also apparent that Ella Hurlstone had at one time been a schoolteacher.

Eccleshall
Stafford
England.
Sept 20th

(1)

Dear Joy
 Now I think I shall have to do some guessing and hope that I guess right. I had a lovely little packet of views the other day from Adelanto California — I know Milton's mother signs herself Ruth — I know Milton has a sister but I didn't know her name. Is that who you are? The views were lovely & serve to show us in England how very different your country is from

(2)

ours & indeed how very different seperate States in America can be. Of course we have never seen dates growing, nor realised exactly how they did grow. We have only bought them before at Xmas time (pre-war only) when we should pay about "a quarter" for a box containing about 36 dates arranged along a stalk, apart from that they can be bought stoned in a solid block & are used to put in puddings to make them a little sweet when we have no sugar.

I should have liked to have sent you some nice views back, but we have no nice ones yet. However, the other day I came across these two little pieces of "Goss" china. They were made in

Staffordshire by (3) a firm a long time ago, — there has been none made for 20 years and they are getting impossible to buy. I was glad to get one for you that had the "coat of arms" of Staffordshire on it, since Milton has spent most of his time in that county. The "arms" show Stafford Castle and two Staffordshire "knots" the symbol of the County — a knot that "holds fast."

The other is just an odd little piece — a model of an old English candle "snuffer". Held down over a lighted candle it puts out the flame by shutting off the air but it does not hurt the wick — as it is pointed. All candlesticks had a "snuffer"

in the days about 1600 A.D to 1700 A.D. (before)
when candles were the only
lights & hundreds had to be
lit & put out in peoples houses.
For the tall candles on walls
& on the big glass 'chandeliers'
a snuffer on a stick was used
like this — one is still used now
every Sunday in our church to
put out the altar candles.
About 200 years ago when
sedan chairs chairs
where used to carry people about,
torches were carried to light the
way through the streets & on the
iron gates of the large houses a
very big 'snuffer' in iron was
placed on either side, for the men
to 'snuff' out the large torches.
We had two fixed on my College

5.

gate which were about 400 years old.

There that is a "lot" about two very little pieces of china. As "china" they are worth looking at — as they are made of very refined clay and if you hold them up to the light you can see the shadow of your finger through them — that is how to tell "china" from "pitcher". This china was too refined & fragile to make household things and was only used for ornaments.

Now I am very sorry but I cannot give us any news of Milton just at present. From what one of the other MP's said I think he must have been away from this camp for a time. I expect he will be calling in to see me some day and I will let your mother know how he is looking.

They are all very busy now at this camp – it is being used to check the men returning to the U.S.A. from Germany. 35,000 went through last month – 70,000 will go through next month & 100,000 in the next month to Xmas. After that we shall not see many more. I am so glad for your mother, that the war is over & Milton safe – I hope it won't be too long before he can come home. Thank you once again for the views

Yours sincerely
Ellen Hurlstone.

5. War Brides

It was, given the proximity and the duration, inevitable that some GIs and British women would fall in love and even sometimes marry each other. From the available statistics about one quarter of a percent of the GIs took wives in the UK, with the vast majority being English. Brief general treatments of the British war brides are found in (Bowman) *Wild Blue Yonder,* (Gardinor) *Overpaid, Oversexed & Overhere* and (Millgate) *Got Any Gum Chum?*. A more detailed history of the British war brides is provided in (Reynolds) *Rich Relations.*

6. Armed Services Books

The books were small enough to slip into a uniform pocket. The books had soft, flexible covers and were printed in small print on thin paper. The books ranged in size, depending on the length of the texts. Typical sizes ranged from a half-inch thick by four inches wide and five and a quarter inches long to three quarters of an inch thick by four and a half inches wide by six and a half inches long. The larger size was about the upper limit of what could be carried in a fatigue uniform pocket. The longer texts were frequently abridged to meet the upper size limit. Subject matter of the books was most of the then current and probably all of the classical fiction and non-fiction literature.

The Armed Services Additions Books were a most extraordinary enterprise by the Council on Books in Wartime. The members of the armed services during World War II were extremely grateful for all of these books. Providing all of the books free to the armed services is something of which the publishers of America should always be proud. It is impossible to know just how many millions of hours of sheer boredom that were ameliorated by all of the issues and volumes of the Armed Services Editions Books.

The following notice appeared on the inside of the front cover of each volume of the Armed Services Additions Books

MILTON COOK

ARMED SERVICES EDITIONS

THIS BOOK is published by Editions for the Armed Services, Inc., a non-profit organization established by the Council on Books in Wartime, which is made up of American publishers of General (Trade) books, librarians, and booksellers. It is intended for exclusive distribution to members of the American Armed Forces and is not to be resold or made available to civilians. In this way the best books of the present and the past are supplied to members of our Armed Forces in small, convenient, and economical form. New titles will be issued regularly. A list of the current group will be found on the inside back cover.

NOTICE

7. Kimbolton Air force Station

    A map of the Kimbolton AAF Station 117 in East Anglia is shown on page 277 of (Freemen) *The Mighty Eighth War Manual*. The details of the units located at The Kimbolton AAF station are available in (Freemen) *The Mighty Eighth*.

8. The B-17 Flying Fortress Bomber

Feature	Parameter
Length	74 ft. 9 in.
Wing Span	103 ft. 10 in.
Height (Vertical Fin)	19 ft.
Weight (Gross)	60,000 lbs.
Bomb Load (Average)	5,000 lbs.
Speed (Cruising)	211 m.p.h.
Horsepower	4800 hp
Armament (.50 cal machine guns)	12
Crew	10 *

　* Later reduced to 9

9. Comment—"in draft" is listed as a synonym for maelstrom (Rodale) *The Synonym Finder.*

10. The V-1 Buzz Bomb

    The V-1, also named the "Doodle Bug," was the first primitive guided missile, a flying bomb. The fact that this particular V-1 buzz bomb had flown this far north was probably due to a failure of the guidance systems mechanical counter. The counter would open a switch after completing a preset countdown that turned off the fuel to the pulse jet engine, at which time the missile would go into a steep dive, exploding on impact. The warhead contained 1,760 pounds of high-blast amatol, an explosive that was more powerful that TNT. The fuel load was 1,333 pounds of "B-stuff," which was a mixture of gasoline and additives to increase power. Dimensions were: overall length 27 feet, 3½ inches; maximum fuselage diameter 2 feet 9 inches; and wingspan 17 feet 4½ inches. The pulse jet engine was mounted on top of the fuselage. The "V" destination stood for "Vergeltungs Waffe" or "vengeance weapon."

11. The B-24 Liberator Bomber

Feature	Parameter
Length	66 ft.
Wing Span	110 ft.
Height (Vertical Fin)	18 ft.
Weight (Gross)	60,000 lbs.
Bomb Load (Average)	8,000 lbs.
Speed (Cruising)	225 mph
Horsepower	4800 hp
Armament (.50 Caliber Machine Guns)	10
Crew	10 *

   * Later reduced to 9

12. The P-38 Lightening Fighter

    The Lockheed P-38 fighter aircraft was not very effective operating as a bomber escort in the very cold 30,000-foot altitude over Europe. The plane did perform very well in the warmer climates of the Mediterranean and Pacific war zones.

Feature	Parameter
Two Allison in line engines	1425 hp
Wing Span	53 ft.
Length	37 ft. 10 in.
Empty Weight	12,380 lb.
Maximum Loaded Weight	20,300 lb.
Maximum Speed	402 mph @ 25,000 ft.
Cruise Speed	250-320 mph @ 30,000 ft.
Normal Range	300 miles
Maximum Range with Drop Tanks	350 miles
Armament	One 20 mm cannon & four @ 50 caliber machine guns

13. Americans in the RAF

(Holmes) *American Eagles, Americans in the RAF 1937-1943*. This book provides an account of the American fighter pilots who flew in the RAF Eagle Squadrons and their subsequent transfer into the U.S. Air Force.

14. The P-61A Black Widow Night Fighter

Feature	Parameter
Length	49 ft. 7in.
Wing Span	66 ft.
Bomb Load	4 @ 1,600 lb.
Speed	375 mph
Horsepower	2,000 per engine,—4,000 total
Armament	4 @ 20 mm cannon, 4 @ .50 cal. Machine guns
Crew	3

15. Clark Gable-Airman

In his wrap up Clark Gable's biographer, Warren G. Harris, concludes that Gable is *the standard against which all other screen actors are measured*. Also, he maintains that as a motion picture actor Gable has never been surpassed in seventy years. But in addition to the unique status of being **the movie actor** Gable also became a historical personage as a soldier; an airman in the Mighty Eighth Air Force. But he was never able to completely disconnect from being **the movie actor.**

He had lost his wife, the love of his life, while she was serving her country. His wife, actress Carole Lombard was killed in an airplane crash while hurrying home from a tour selling war bonds. Gable, as vice chairman of the Hollywood Victory Committee, had arranged for Lombard to take part in the tour. On August 12, 1942, at the age of 41, he enlisted in the Army Air Force as a private; Lombard had wanted him to enter the military service. At the time he signed up he was making $357,000, the second-highest salary in the United States. With his public stature, he could have requested and receive a direct commission as an officer of in some special services function. But instead, he chose to do his duty in just the same manner as all of the rest of us did when called to serve.

While we were undergoing a basic training in Miami Beach, I saw in one of the "base" motion picture theaters, as part of the regular program, a short documentary of Gables' transition through the Miami Beach Officers Candidate School. After his graduation from OCS Gable met privately with General H. H. "Hap" Arnold, who had presided at the commencement, to discuss the future. As a result of this meeting Gable was to attend the Air Force gunnery school and then work on a aerial gunner recruiting film in the Documentary Field Unit of the Eighth Air Force. This private meeting of a newly commission Second Lieutenant with the Commanding General of the Army Air Force was very unusual. But General Arnold had a significant problem of just how to fit Gable into Air Force operations. Unlike say the seven years younger James Stewart, who was a licensed aircraft pilot with 400 hours in his logbook; Gable did not have any outstanding skills that were readily transferable to military operations. (Jimmy Stewart would fly 20 missions as a bomber pilot in the Mighty Eight.) So the aerial gunner recruiting film was some sort of a make-work assignment for Gable, who because of his prominence could not easily be refused a combat assignment. The premise that the recruiting film was most likely a make-work assignment is based on the fact as there was already a good recruiting program in place known as "the draft".

In January 1943, days after his 42nd birthday Gable finished training and received his silver aerial gunner wings. He then shipped out to

England to his assignment with the 351st Bomber Group, which was flying B-17 Bombers. The plane in which he was flying as a combat photographer was hit on each of his five missions. The Air Force probably did not want to expose Gable to anymore combat because as some sources suggest, Field Marshal Hermann Goring had offered a prize for any member of the Luftwaffe who could bring Gable down. In any case the odds would become worse for being hit on each additional mission on which Gable flew. At the completion of his combat service Gable was awarded the Distinguished Flying Cross and the Air Medal. (At this point in time the normal tour of duty in the Eighth Air Force was twenty-five bomber missions. Each airman was awarded the Air Medal for every five missions completed and a DFC at the completion of twenty-five missions. While this might appear to be a somewhat inordinate apportionment of decorations for extraordinary achievement involving combat it was justified by the circumstances. Flying at 30,000 feet in slow un-pressurized bombers at 60 below temperatures through heavy flak over target areas and sustained attacks by Luftwaffe fighters elsewhere, the Might Eighth sustained casualties at a rate comparable to that of an infantry unit. The Eighth Air Force was one out the most highly decorated units in the Second World War, including 6,845 Purple Hearts plus 188 Oak Leaf Clusters for a total of 7,033.)

No doubt because of this prominence Gable's presence in England was frequently noted and recorded. It is almost a requirement that any history dealing with the US Army Air Force in Britain during World War II at least mention Gable, as so have I. There are many vignettes, passing references and sometimes at the very least a picture of Gable in the following (and probably many more) publications: (Astor) *The Mighty Eighth*, (Anderson} The *Men of the Mighty Eighth*, (Bodie, Macpherson) *America's Mighty Air Force*, (Bowman) *Castles in the Air*, (Freeman) *The Mighty Eighth*, (Gardiner) *Overpaid, Oversexed & Over Here*, (Millgate) *Got Any Gum Chum?* and (Reynolds) *Rich Relations*.

Gable returned to the United States in October 1943 with orders to stop in Washington, DC to confer or with General Arnold. As it turned out Arnold had forgotten why he sent Gable to England. Gable told Arnold, "I was too make a film to help recruit gunners." To which Arnold replied that we've licked that problem. Informed that 50,000 feet of film (in color) had been shot Arnold told Gable "do anything you want, I'm sure you can make a documentary that will do the Air Force proud".

Stationed at the Hal Roach Studios in Culver City, Gable cut the documentary *Combat America* from the 50,000 feet of film. *Combat America* is a somewhat unsophisticated war propaganda film narrated by Gable himself. A point of interest is a short scene in which Gable talks with the copilot who is recovering from a wound. In the discussion of this incident they note that the guns had been stowed and then just as they reached the coast of England their bomber was under attack by a German fighter aircraft. This same event is described in (Bowman) *Castles in the Air.*

16. Promotion Mystery

Here is a mystery, On the Special Orders dated 10 March 1944, I was still a Private. On the Special Orders dated 6 January 1945, I was now a Private First Class. I was promoted some time in the interim, but it was apparently not recorded in my service record. My discharge from the Army states that I was promoted to PFC on arrival in the ETO. If this were in fact the case, the Army would have owed me some back pay. What I think occurred was that the Company Clerk was pressed into service, with the station personnel processing paperwork for the soldiers passing through the Repple Depple. While he was so employed, he did not have the time or the interest to maintain our service records. I think that this was the case as the Repple Depple received a Meritorious Unit award. And also, this award was not entered on my discharge. The Meritorious Unit award was a two-inch square OD patch with a gold laurel wreath, which was worn on the lower right sleeve of the uniform jacket.

17. The V-2 Rocket

The V-2 was the first practical application of the liquid fueled rocket technology developed first by Robert Goddard in 1926. The Germans could not obtain information from Goddard and so developed liquid fueled rocket technology a second time in 1930-31. In both cases, space travel was the primary objective.

The nose cone of the V-2 rocket contained a one-ton, explosive-filled warhead. Flight control technology consisted of an automatic gyroscopic guidance package and guidance beam and radio control receivers. One rocket fuel tank was filled with an alcohol-water mixture, and the other with pressurized oxygen. A small tank of peroxide was used to generate steam to drive the fuel pumps. The rocket engine developed 56,000 pounds of thrust. Air dynamic flight control was provided by rudders, which were located on the outer edges of the four fins. Graphite vanes in the rocket engine exhaust stream provided dynamic control in space.

Feature	Parameter
Overall Length	46 ft. 5 1/16 in.
Maximum Diameter	5 ft. 4 7/8 in.
Fin Span	11ft. 8 5/16 in.

In 1947, while at the Naval Research Laboratory in Washington, D.C., I had the opportunity to walk inside of the empty fuselage of a V-2 resting on its side. This is quite possibly the same V-2 fuselage that is now on display in the Smithsonian Air & Space Museum. The V-2 airframe was typical aircraft stressed skin, longeron and frame type construction.

18. The P-47 Thunderbolt Fighter

The Republic P-47 Thunderbolt Fighter aircraft, using the most powerful turbo-charged radial engine available was designed as a high altitude interceptor. It was used in the Eighth Air Force as an escort fighter for ground attack and troop support. The airplane was larger and about twice the weight of most of the other fighters used in World War II and could out dive them all.

Feature	Parameter
Engine (without water injection)	2,000 hp
Engine (with water injection)	2,300 hp
Wing span	40 ft. 9 in.
Length	36 ft. 1 in.
Empty weight	9,900 lb.
Maximum loaded weight	15,000 lb.
Maximum speed	433 mph @ 30,000 ft.
Cruise speed	275 mph
Normal range	250 miles
Maximum range with drop tanks	475 miles
Armament	eight .50 caliber machine guns

19. Evidence of the sometimes-farcical essence of the United States Navy
The following collection of observations that have aggregated on the Internet, hold at worst a sometimes only slightly warped mirror up to the nature of life in the U.S. Navy.

How to stimulate the life of a sailor:

- Buy a steel dumpster, paint it gray inside and out, and live in it for six months. Run all the pipes and wires in your house exposed on the walls.
- Repaint your entire house every month.
- Renovate your bathroom. Build a wall across the middle of the bathtub and move the shower—head to chest level. When you take showers, make sure you turn off the water when you soap down
- Raise the thresholds and lower the headers of your front and back doors so that you either trip or bank your head every time you pass through them.
- Disassemble and inspect your lawnmower every week.
- On Mondays, Wednesdays and Fridays turn your water heater temperatures up to 200°. On Tuesdays and Thursdays, turn the water heater off. On Saturdays and Sundays tell your family they used too much water during the week, so no bathing will be allowed.
- Raise your bed to within 6 inches of the ceiling, so that you cannot turn over without getting out and then getting back in.
- Sleep on the shelf in your closet. Replace the closet door with a curtain. And then have their spouses whip opened the curtain about three hours after you go to sleep, and then shine a flashlight in your eyes and say "sorry, wrong rack."
- Have your family qualify to operate each appliance in your house: dishwasher operator, blender technician, etc.
- Have your neighbor come over each day at 5 a.m., blow whistle loudly, and shout "reveille, reveille, all hands heave out and trice up."
- Have your mother-in-law write-down everything she's going to do the following day, and then have her stand in your backyard at 6 a.m. while she reads it to you.
- Submit a request chit to your father-in-law requesting permission to leave your house before 3 p.m.
- Empty all the garbage bins in your house and sweep the driveway three times a day, whether it needs its or not.

MILTON COOK

- Have your neighbor collect all your mail for a month, read your magazines, and randomly lose every fifth item before delivering it to you.
- Watch no TV except for movies played in the middle of the night. Have your family vote on which movie to see, and then show a different one.
- Make your family menu week ahead of time without consulting the pantry or the refrigerator.
- Post a menu on the kitchen door informing your family that they are haveing steak for dinner. Then make them wait in line for an hour. When they finally get to the kitchen, tell them you are out of steak, but they get to have dried ham or hot dogs. Repeat daily until they ignore the menu and just ask for hot dogs.
- Bake a cake. Prop up one side of the pan so the cake bakes unevenly. Spread icing real thick to level it off.
- Get up every night around midnight and have a peanut butter and jelly sandwich on stale bread. (Midrats)
- Set your alarm clock to go off at random times during the night. At the alarm, jump up and dress as fast as you can, make sure to let in your top shirt button and tuck your pants into your socks. Round out into the backyard and uncoil the garden hose.
- Every week or so, throw your dog in the pool and shout, "Man overboard portside!" Rate your family members on how fast they respond.
- Place a podium at the end of your driveway. And have your family stand watches at the podium, rotating at four-hour intervals. This is best done when the weather is worse. January is a good time.
- When there is a thunderstorm in your area, get a wobbly rocking chair, said in it and rock as hard as you can until you become nauseous. Make sure to have a supply of stale crackers in your shirt pocket.
- Make coffee using 18 scoops of budget priced coffee grounds per-pot; allow the pot to simmer for five hours before drinking.
- Have someone under the age of ten give you a haircut with sheep shears.
- Sew the back pockets of your jeans on the front.
- Lock yourself and your family in the house for six weeks. Tell them that at the end of the sixth week you were going to take them to Disney World for "liberty." At the end out of the sixth week informed

them that the trip to Disney World has been canceled because they need to get ready for an inspection, it will be another week before they can leave the house.

q.e.d.

20. Comments on the P-Kit.

On page 65 of *Wild Blue Yonder* (Bouman) has included photographs of a P-Kit and a C-Device. The OD P-Kit is a Museum quality historical artifact that someone had sagacity to save. On the other hand the C-Device is not government issue and in all probability no one had the foresight to save one for the perusal of posterity. By using a magnifying lens I was able to clearly read the contents of the P-Kit package. It is good that Bouman provided such a clear picture of the P-Kit because how would one know where to look to find this singular information.

Appendix I

Special Orders and Travel Documents

1. Special Orders No. 70, Dated 10 March 1944.

2. Train Schedule for Special Orders No. 70.

3. Special Orders No. 6, Dated 6 Jan 1945.

4. Train Schedule for Special Orders No. 6.

5. Special Orders No. 66, Dated 7 Mar 1945.

6. Special Orders No. 78, Dated 19 Mar 1945.

7. Train Schedule for Special Orders No. 78.

8. Special Orders No.82, Dated 23 Mar 1945.

9. Security Clearance for Special Orders No. 82, Dated Mar 29, 1945.

10. Special Orders No. 38, Dated 30 Aug 1945.

11. Special Orders No. 68, Dated 29 Sept 1945.

RESTRICTED

HEADQUARTERS
EIGHTH AIR FORCE REPLACEMENT DEPOT
APO 635

SPECIAL ORDERS)
)
No.   70   )   EXTRACT

APO 635
10 March 1944

SECTION I.

* * *

14. 2nd Lt/Mr. J. Cook XXXXXXX (O??) Hq 92nd Stn Comp Sq (X), AAF Sta 595, WP o/a ABT o/a 11 March 1944 on TD for two (2) days to AAF Sta 117, for the purpose of escorting 1st Lt XXXXXX A. XXXXXXX XXXXXX (O11) AAF Sta 14th AAF, this stn, now in confinement that stn, to confinement AAF Sta 595, reporting on arrival to CO thereat. If government mess and qrs are not available for the period while on travel c/o CO status the EM will pay 75¢ each The mem allow in lieu of rat and qrs in accordance with policy 35-1 Hq XXXXX, UK, 5-11-43, and evening rat in kind will be provided XXX/Delays while traveling. The cost of trans and meals will be charged against 1st Lt/Mr.XXXX in accordance with Par 21b AR 35-120 and Par 14, AR 35-4520. CPbk. QMC. SNY. FD:1 91-1 P 432-02 A 212/40420.
AUTH Ltr Hq, XXXXXX, 5-11-43.

* * *

By order of Col XXXXX:

                                        ROBERT W. MOORE,
                                        Capt AC,
                                        Acting Adjutant General.

OFFICIAL:

ROBERT W. MOORE,
Capt AC,
Acting Adjutant General.

- 1 -

RESTRICTED

TRAIN SCHEDULE

OFFICER IN CHG: Stowe FOR: 2ILR DATE: 7-44
SPECIAL TRAIN NO: 70
RUN: 18 OFF in charge: A. M. Cook
                         9456 4984

LOCATION	ARR. TIME	DEPART TIME	REMARKS
Stoke	0740	0744	
Derby	0856	1058	
Kettering	1816	1810	
Kimbolton	1887		

1. Person in charge will notify all in his party of their final call in line list before they entrain.
2. If additional assistance is required, contact the local train, either British or American.
3. All personnel are responsible for the loading and transfer of their own baggage.
4. When transfer of trains is necessary, plan to do so promptly and secure baggage immediately.

RESTRICTED

Headquarters
70TH REPLACEMENT DEPOT (AAF)
APO 558

AAF Sta 594.
6 Jan 1945.

SPECIAL ORDERS)
            :
NO.   6)   E X T R A C T

SECTION I

*   *   *

9. Pfc Milton E. Cook 35584484 (677) CMP, AAF Sta 594, WP m/o CMP o/a 6 Jan 1945 on TD not to exceed two (2) days to London for the purpose of escorting Pvt James E. Halerich 35062196 (021) AC, Casual Pool, this Hq, atchd 154th Replacement Co, 129th Replacement Bn (AAF), this sta, fr confinement Military Authorities, London, to confinement this sta. rptg on arrival to CFO thereof. Reimbursement for qrs and rat is auzd Pfc Cook while in a travel m/o TD status in accordance w/Cir 62, Hq European T of Opns, USA, 5-6-44 as amended by Cir 84, Hq European T of Opns, USA, 31-7-44. Cooked rat in kind will be provided Pvt Halerich while traveling. The cost of trans and subs will be charged against Pvt Halerich in accordance with Par 23b, AR 35-120, and Par 14, AR 35-4520. CTNS. TCNT. TDN. FD& 50-136 P 432-02 A 212/50425. AUTH: Ltr, Hq European T of Opns, USA, AG 300.4, 22-11-44.

*   *   *

By order of Colonel BARHN:

                                        WILLIAM A. MCKEOWAN
                                        Major, AGD
                                        Adjutant General

OFFICIAL:

WILLIAM A. MCKEOWAN
Major, AGD
Adjutant General

- 1 -

RESTRICTED

TRAIN SCHEDULE

STATE OF EM: _Leave_ TIME: _1654_ DATE: _____
SPECIAL TRAIN NO: _____ Off. in Charge: _____
FROM: _____

LOCATION	ARRIVE	DEPART	REMARKS
Stafford	1718	1748	Change
London	1858		

1. Person in charge will notify all in his party of tentative
   ... destination before they arrive.
2. If additional assistance is required, contact the local
   ... either British or American.
3. All personnel are responsible for the loading and transfer
   of luggage and baggage.
4. When transfer of trains is necessary, plan to detrain
   promptly and assemble in station.
5. The attached key Deposit or Railway Warrant is to be
   given the ticket collector at ultimate destination.

Form No. EPD-101

RESTRICTED
HEADQUARTERS
70TH REINFORCEMENT DEPOT (AAF)
APO 562

AAF STA 594,
7 Mar 1945.

SPECIAL ORDERS)
NO.   66)   EXTRACT

* * *

7. Pfc Milton H. Cook 39884494 (Svc) Cur 334 Stn Com Sq Replacement Depot (Op) this sta dy WP AAF Sta 590 o/a 7 Mar 45 on TDY not exceed two (2) days purpose escorting Pvt Frank L. Biossiat 38150462 (SSO) AC Casual fmd this Hq reld atchd 126th Reinforcement to 125th Reinforcement Bn (AAF) this sta and trfd in gr Stn Com Sq Base Air Depot Area No 1 AAF Sta 590 fr confinement this sta to confinement AAF Sta 590 reporting on arrival to CO. Reimbursement for qrs and rat is auth Pfc Cook while in a Tvl and/or TDY status in accordance w/Cir 63 Hq European TO, USA 5 Jun 44 as amended by Cir 54 Hq European TO, USA 31 Jul 44. Rat in kind will be provided Pvt Biossaat while traveling. Tvl by ry and govt MT is auth. Upon compl of TDY Pfc Cook will ret AAF Sta 594. TCNT. TDN. FD* 60-136 P 433-02 A 212/50425. Auth: Ltr Hq European TO, USA AG 300.4 22 Nov 44 and secret Ltr file 252 Hq Base Air Depot Area AAF US Strategic Air Forces in Europe 7 Nov 44.

By order of Colonel RAMEY:

                              E. J. MANY
                              Captain, AC
                              Actg Asst Adj Gen

OFFICIAL:

E. J. MANY
Captain, AC
Actg Asst Adj Gen

RESTRICTED

RESTRICTED

HEADQUARTERS
70TH REINFORCEMENT DEPOT (AAF)
APO 559

AAF STA 594.
19 Mar 1945.

SPECIAL ORDERS
NUMBER   78   EXTRACT

*   *   *

23. Pfc Milton F. Cook 30664484 (677) AC Hq &ta Com Sq, Reinforcement Depot
(AF) this sta WP to 112th Reinforcement Bn (AAF) AAF sta 163 o/a 20 Mar 45 on TDY
not exceed three (3) days purpose escorting Pvt Roger B. Church 31675030 AC Casual
fr this Hq stand from Reinforcement Co 112th Reinforcement Bn (AAF) AAF Sta 163
fr confinement this sta to confinement ANY Hq 163 reporting on arrival O.I. Reinforcement for qrs and rat auth Pfc Cook while in trvl and/or TDY status in accordance w/Cir 63 Hq European TO, USA & Cir 44 as amended by Ltr 84 Hq European TO,
USA 21 Jul 44. Rat in kind will be provided Pvt Church while traveling. Trvl by
MT and govt MT is auth. Upon compl TDY Pfc Cook will ret proper sta. TDN. TBMA.
FD 60-136, P 432-02 A 212/50425. Auth: UP Hq European TO, USA AG 200.4 WTX
24 Nov 44.

*

BY ORDER OF COLONEL BAKER:

OFFICIAL: R. J. HART
                                 Captain, AC
                                 Actg Asst Adj Gen

R. J. HART
Captain, AC
Actg Asst Adj Gen

RESTRICTED

TRAIN SCHEDULE

STATION OF DES: *Stone* TIME: *1001* DATE: _____
SPECIAL ORDER NO: _____ Off in Charge: _____
PAR: _____

LOCATION	ARR. TIME	DEPART TIME	REMARKS
Stoke	1017	1110	Change
London	1410	—	Euston Station
"	—	1342	L'pool Station
Stansted	1648		

1. Person in charge will notify all in his party of the final Rail destination before they entrain.
2. If additional assistance is required, contact the local R.T.O., either British or American.
3. All personnel are responsible for the loading and transfer of their own baggage.
4. When transfer of trains is necessary, plan to detrain promptly and secure baggage immediately.
5. The attached War Department Railway Warrant is to be given the ticket collector at ultimate destination.

70th RD RTO - 129

RESTRICTED
HEADQUARTERS
70TH REINFORCEMENT DEPOT (AAF)
APO 652

AAF Sta 594.
23 Mar 1945.

SPECIAL ORDERS
NUMBER 82

EXTRACT

x   x   x   x

17. Sgt Donald R. Foiles 39836447 (677) AC and Pfc Milton E. Cook 39864484 (677) AC 93d Sta Com Sq Reinforcement Depot (Sp) this sta WP Hq 134th Reinforcement Bn(AAF) AAF Sta 392 o/a 23 Mar 45 on TDY not exceed five (5) days purpose escorting fol AC EM orgns indicated fr confinement this sta to confinement AAF Sta 398 to await transportation to proper orgn reporting on arrival CO. Reimbursement for rat and qrs auth Sgt Foiles and Pfc Cook while in a tvl and/or TDY status in accordance w/Cir 63 Hq European TO, USA 5 Jun 44 as amended by Cir 84 Hq European TO, USA 31 Jul 44 and in accordance w/existing laws and regulations and rat in kind will be provided fol EM while traveling. Tvl by ry, govt MT and mil acft is auth. Upon compl TDY Sgt Foiles and Pfc Cook will ret proper sta. TCNT. TDN. FSA 60-136 114 ? 432-02 A 212/50425. Auth: Ltr Hq European TO, USA AG 300.4 AFM 22 Nov 44 Cir 120 Hq European TO, USA 13 Dec 44 and USSTAF Reg 80-14, 5 Mar 45.

                                          WARMINGBORGH
Tec 5 Edward E. Webster 35864771   82d Sv Sq
S Sgt Barney E. Cash 14069499   587th Bomb Sq 394th Bomb Gp
                                          APO 140

x   x   x   x

BY ORDER OF COLONEL RADER:

OFFICIAL: E. J. HART
                                 Captain, AC
/s/ E. J. Hart   Actg Asst Adj Gen
/t/ E. J. HART
    Captain, AC
    Actg Asst Adj Gen

DISTRIBUTION
    A

A TRUE COPY

/s/ Joseph M. Rogers
JOSEPH M. ROGERS
1st Lt., Air Corps
Adjutant

RESTRICTED

AAF STATION 180
502D TRANSPORT WING
SECURITY CONTROL OFFICE

DESTINATION U.K.　　　　　　　　　　DATE ~~MAR 25 1945~~

NAME Cook, M.E.　　　　RANK Pfc　ASN 39564484

THE ABOVE-NAMED PERSON HAS BEEN CLEARED THROUGH THIS STATION AND HAS
SIGNED A DECLARATION TO THE EFFECT THAT HE/SHE DOES NOT HAVE ANY UN-
CENSORED OR UNAUTHORIZED CLASSIFIED MATERIAL IN HIS/HER POSSESSION.
HE/SHE HAS BEEN IDENTIFIED TO THE SATISFACTION OF THE COMMANDING
OFFICER AND IS IN POSSESSION OF COMPETENT ORDERS.

　　　　　　　　　　　　　　　　　Clayton E. Fenton
　　　　　　　　　　　　　　　　　CLAYTON E. FENTON
　　　　　　　　　　　　　　　　　1ST LT., AC,
　　　　　　　　　　　　　　　　　SECURITY CONTROL OFFICER

NOTE: TO BE SHOWN TO THE TRAFFIC OFFICER UPON BOARDING A/C AND
SURRENDERED TO SECURITY CONTROL OFFICER AT DESTINATION. (VALID ONLY
ON DATE STAMPED ABOVE)

PCI FORM NO. 2.

　　　　　　　　　RESTRICTED

R E S T R I C T E D

HEADQUARTERS
AAF/BT REINFORCEMENT DEPOT (PROV) (MAIN)
APO 652

SPECIAL ORDERS)   AAF STA 594.
NUMBER 38)   EXTRACT   30 Aug 1945.

*   *   *

11. Pfc Milton E. Cook 39564684 (67) 470 93rd Sta Com Sq Reinforcement Depot (Sp) this sta WP APO 413 o/a 31 Aug 45 purpose escorting Pvt James J. Slattery Jr 11046111 AC Det "11" Sta Com Sq Base Air Depot Area No 2 APO 635 fr conf APO 413 to conf this sta. Reimbursement for qrs and rat auth Pfc Cook while in a tvl and/or TDY status in accordance w/Cir 32 Hq UK Base, 15 Jul 45. Reimbursement for rat is not auth for any period of TDY in the London area. Reimbursement for qrs during period of TDY in London area auth only upon certification of Billeting Officer UK Base 5 N. Audley St, London that govt qrs were not available. EM concerned will report to Billeting Officer S N. Audley St London for assmt to billets. Rat in kind will be provided Pvt Slattery Jr while traveling. Tvl by ry and govt MT is auth. Upon compl TDY Pfc Cook will ret proper sta. TCNT. TDN. FSA 60-136 P 432-02 A 212/60425. Auth: Ltr Hq European TO, USA AG 300.4 SPN 22 Nov 44.

*   *   *

BY ORDER OF COLONEL CLYDE:

OFFICIAL: DOROTHY A. KIMMEL
                                                                     1st Lt,   WAC
                                                                       Actg Asst Adj Gen

DOROTHY A. KIMMEL
1st Lt,   WAC
Actg Asst Adj Gen

R E S T R I C T E D

HEADQUARTERS
AAF/AF REINFORCEMENT DEPOT (PROV) (MAIN)
APO 595

AAF STA 594.
SPECIAL ORDERS   20 Sep 1945.
NUMBER   68   EXTRACT

* * *

8. Pfc Milton E. Cook 32654454 (SFF) AC HQ AAF Sta 594 WP APO 413 o/a 20 Sep 45 on DY not exceed two (2) days purpose escorting Pvt Edwin I. Jones 37035654 (WS and ASN unknown) Hq Co AAF APO 755 fr conf APO 413 to conf while en r. Reimbursement for qrs and rat auth Pfc Cook while in tvl status in accordance w/Cir 51 WD Hac 15 Jul 44. Reimbursement qr rat is not auth for any period of TDY in the London area. Reimbursement for qrs during period of TDY in London area auth only upon certification of Billeting Officer HQ Base S L Audley St London that govt qrs were not available. EM concerned will report to Billeting Officer S L Audley St London for asgnt to billets. Rat in kind will be provided Pvt Jones while traveling. Tvl by ry and govt MT is auth. Upon compl TDY will ret proper sta. TDN. TBR. FD CO-1367 / OSS-CS A 212/50425. Auth: Ltr Hq European TO, USA AG 200.4 MPM 22 Nov 44 and Cir 108 Hq US Forces European Theater 15 Jul 45.

* * *

BY ORDER OF COLONEL KADEL:

OFFICIAL: DOROTHY A. KIMMEL
                                    1st Lt   WAC
                                    Actg Asst Adj Gen

DOROTHY A. KIMMEL
1st Lt   WAC
Actg Asst Adj Gen

- 1 -
R E S T R I C T E D

Appendix II

Other Documents

This final collection of documents, five passes and a PX ration card, also survived in my papers. It would have been somewhat tedious to include them in the main portion of this memoir. But they do provide some additional information on how the war was run; so here they are along with some pertinent comments. One apparent thing is that these and most of the other included documents appear to be consistently changing content and format. This organized muddling is perhaps a metaphor for the conduct of WW II; keep changing and trying new things and hopefully sometimes things will work out just right.

```
    CLASS "B" PASS 1228 M.P. Co. (Avn)
    Milton E. Cook   Pvt. 39564484
                name and grade              ASN
     18       160      6'0"    Brown     Blue
     age     weight   height  color of hair  color of eyes
    Is authorized to be absent from his station
    when not on duty from retreat to 11:30 P.M. on
    Sundays and week days and until 12:30 Sun-
    day A.M. on Saturdays.
```

CLASS "B" PASS

My first class "B" pass. The Station in this case was one floor of a high-end hotel situated by the ocean in North Miami Beach.

```
                MESS PASS
    1257th M.P.

    Cook, Milton E                    Pvt

    Bkf        =        0700 hrs.
    Din        =        1100 hrs.
    Sup        =        1700 hrs.

    Mess Officer  [signature]
```

MESS PASS

These two passes are the only documents that actually specify the unit number of the Military Police Company. All other documents identify a higher-level unit to which the MP Company was assigned.

A TWO-DAY PASS

    This pass is for the 28th through the 30th of December 1943, but if not carefully read it appears to be for the year 1942. At this time Chorley (Washington Hall) was part of the 12th Replacement Control Depot. This pass is unique in that it is the only one that I saved which does not have some requirements printed on its revered side.

A ONE-DAY PASS

This is an over night pass from the afternoon of the 10th of December 1943 to the next afternoon. This may be my first pass after our return from TDY at Chorley. This pass is on a standard War Department form that was used prior to issue of the locally generated monthly class "B" passes. Note the oath that had to be signed to obtain the pass. When on duty as MPs we were charged with assuring that other soldiers adhered to these requirements, but when we were off duty it did not matter. This pass is unique in that it is the only one I saved that does not identify my assigned unit.

LAST PX RATION CARD

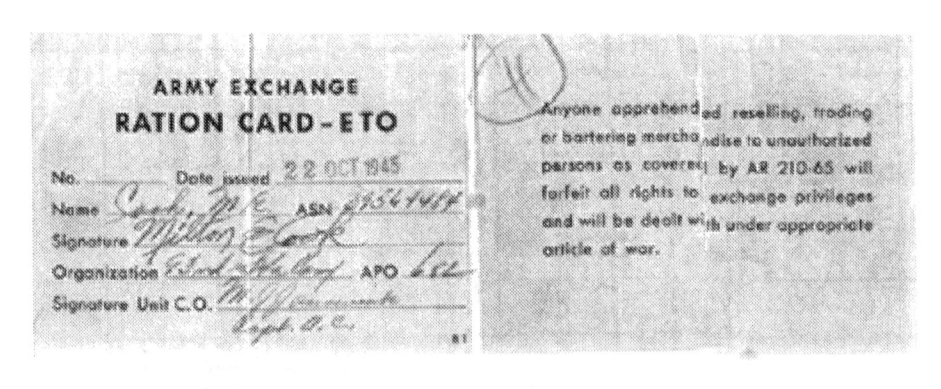

    This later type of ration card probable was re-dated because cards with later dates had not been printed. The reduced number of US Troops still in England at the end of August 1945 probably did not warrant the printing of new PX ration cards. A point of interest is that toothpaste and shaving cream are now rationed items. (Interestingly "tooth paste" has now become one word.)

    I purchased a roll of film on my last visit to the Duncan Hall PX. Used cameras were available but film for them was hard to come by. Any pictures that I may have taken were lost when my camera was stolen from my barracks bag, while it was stacked in the vestibule of our passenger car during our trip from Paris to Munich.

CLASS "B" PASS

This is the last pass that I was issued in England. We would be in Germany before this pass would expire at the end of November 1945. Also, this pass is the only record of our "new" assignment to the reinforcements units.

BIBLIOGRAPHY

Ammer, Cristine, *The American Heritage Dictionary of Idioms,* Boston New York, Houghton Mifflin Co., 2003.
Anderson, Christopher J., *The Men of The Mighty Eighth,* Mechanicsburg, Stackpole Books, 2001.
Astor, Gerald, *The Mighty Eighth,* New York, Dell Publishing, 1998.
Bacon, Benjamin W., *Sinews of War,* Novato, Presidio Press, 1997.
Bodie, Warren M. & Macpherson, Allen, *America's Mighty Eighth Air Force Conception to D-Day,* Haysville, Widewing Publications, 2002.
Bowman, Martin W., *Castles in The Air*, Dulles, Brassey's Inc., 2000.
Bowman, Martin W., *Wild Blue Yonder,* London, Cassell, 2003.
D'Esta, Carlo, *Eisenhower,* New York, Henry Holt and Co., 2002.
Freeman, Roger A., *The Mighty Eighth,* London, Cassell, 2000.
Freeman, Roger A., *The Mighty Eighth War Manual,* London, Cassell, 2001.
Gardiner, Juliet, *Overpaid, Oversexed & Over Here*, New York, Canopy Books, 1992.
Harris, Warren G., *Clark Gable,* New York, Harmony Book, 2002.
Holmes, Tony, *American Eagles,* Crowboruogh, Classic Publications, 2001.
Huie, William Bradford, *The Execution of Private Slovik,* New York, New American Library, Inc., 1954.
King, Benjamin & Kutta, Timothy, *Impact,* Rockville Centre, Sarpedon, 1998.
Millgate, Helen D., *Got Any Gum Chum?*, Thrupp, Sutton Publishing Limited, 2001.
Perret, Geoffrey, *Eisenhower*, Holbrook, Adams Media Corp, 1999.
Pitt, Barrie and Frances, *The* Chronological *Atlas of World War II,* London, Macmillan, 1989.
Reynolds, David, *Rich Relations,* London, Phoenix Press, 2000.
Rodale, Jerome I., *The Synonym Finder,* Emmals, Rodale Press, 1978.

Stubbs, Major David (ed), *The Yarnfield Yank,* Stone, U.S. Army Air Force, 1945.

Taylon, Michael J. H. (ed), *Jane's Encyclopedia of Aviation,* New York, Cresset Book, 1993.

Newspaper, *ARDEE News—Vol.1 No. 1 through No. 11,* Stone, U.S. Army Air Force, 1945.

Report (U.S. Army Air Force), *Target Germany,* New York, Simon and Schuster, 1943.

Ingram Content Group UK Ltd.
Milton Keynes UK
UKHW010701020523
421098UK00001B/50